高职高专电子信息类专业"十三五"课改规划教材

# 计算机专业英语

主编 权小红 李琳 易琼

西安电子科技大学出版社

# 内 容 简 介

目前，计算机技术中直接采用英文的现象越来越普遍，计算机从业者的英文水平对其工作能力起着至关重要的作用。本书即根据高职高专学生的特点为计算机专业学生而编写。

本书共有十个单元，围绕"理论"及"实践"两条主线展开。每一单元的 TEXT、EXERCISES、SUPPLEMENTARY 部分构成了本书的"理论主线"，系统介绍了计算机专业各方向的基础知识及发展热点；CONVERSATION、WRITING 部分则构成了本书的"实践主线"，以计算机英语在工作和生活中的各种典型应用为背景，从口语和写作两方面训练学生的语言实践技能。

本书可作为高职高专院校的计算机专业英语教材、IT 职业英语考试的参考教材或各种相关培训班的专业英语教材，也可供计算机工程技术人员或广大英语爱好者使用。

**图书在版编目(CIP)数据**

计算机专业英语/权小红，李琳，易琼主编. —西安：西安电子科技大学出版社，2012.8(2016.2 重印)
高职高专电子信息类"十三五"课改规划教材
ISBN 978−7−5606−2868−4

Ⅰ. ① 计… Ⅱ. ① 权… ② 李… ③ 易… Ⅲ. ① 电子计算机—英语—高等职业教育—教材
Ⅳ. ① TP3

**中国版本图书馆 CIP 数据核字(2012)第 160402 号**

| | |
|---|---|
| 策　　划 | 邵汉平 |
| 责任编辑 | 雷鸿俊　邵汉平 |
| 出版发行 | 西安电子科技大学出版社(西安市太白南路 2 号) |
| 电　　话 | (029)88242885　88201467　　邮　编　710071 |
| 网　　址 | www.xduph.com　　　　电子邮箱　xdupfxb001@163.com |
| 经　　销 | 新华书店 |
| 印刷单位 | 陕西大江印务有限公司 |
| 版　　次 | 2012 年 8 月第 1 版　　2016 年 2 月第 2 次印刷 |
| 开　　本 | 787 毫米×1092 毫米　1/16　印　张　13.5 |
| 字　　数 | 315 千字 |
| 印　　数 | 3001～6000 册 |
| 定　　价 | 25.00 元 |

ISBN 978−7−5606−2868−4/TP · 1356

XDUP 3160001−2

\*\*\*如有印装问题可调换\*\*\*

本社图书封面为激光防伪覆膜，谨防盗版。

# 前　言

➢ **关于本书**

本书根据全国高等职业教育"十二五"规划教材的指导精神编写。

计算机技术的飞速发展使得我们的地球越来越小，大量新概念、新术语、新资料源源不断地从国外引入，直接采用英文的现象越来越普遍；计算机操作过程中的界面也多是英文。因此，计算机从业者的专业英语水平对其工作能力起着至关重要的作用。

本书参考了大量国内外计算机英文资料和计算机专业英语书籍，根据高职高专学生的认知水平及教学特点而编写，旨在将英语与计算机知识相结合，提高学生独立阅读计算机专业英文资料的能力，强化学生英语实践技能。

➢ **本书结构**

本书共有十个单元，每一单元由以下六部分组成：

- TOPICS：列出本单元的知识技能目标与要点。
- TEXT：本单元的主题文章，概要介绍与本单元主题相关的专业基础知识。每一单元围绕一个主题，涉及计算机软硬件、网络、数据库、信息安全、多媒体等计算机技术的各个方面。
- EXERCISES：通过各种形式的综合练习，加深对本单元主题相关专业知识的理解与掌握。
- SUPPLEMENTARY：拓展阅读，将专业知识与生活实际相联系，就本单元主题，给出生活中一些比较有趣而常见的实际应用。
- CONVERSATION：职业场景对话，每一单元设定一个不同场景，如面试、就餐、招待客户、商业谈判等。
- WRITING：英文应用文写作，每一单元围绕一项应用，如简历、商务合同、通知及邀请函、道歉信及投诉信的写作等。

➢ **本书特点**

Ⅰ.兼顾知识的系统性与技能的实用性。

本书围绕两条主线展开：TEXT、EXERCISES、SUPPLEMENTARY 部分构成了本书的"理论主线"，系统介绍了计算机专业各方向的基础知识及发展热点；CONVERSATION、WRITING 部分则构成了本书的"实践主线"，以计算机英语在工作和生活中的各种典型应用为背景，从口语和写作两方面训练学生的语言实践技能。

Ⅱ.内容选材新颖，难度适中，风格多样。

本书素材均选取计算机专业各个方向的最新信息，内容与时俱进；充分考虑高职高专学生的认知水平，选材难度适中，篇幅合理，通俗易懂；在文体风格上有专业论文、技术手册、科普短文等各种文体形式。

Ⅲ. 注重学生创新性学习能力的培养。

在 EXERCISES 部分布置有各种形式的专业知识练习，可让学生做探究性自主学习；在 CONVERSATION 和 WRITING 部分通过 Practice 环节，布置实践技能开放式训练任务，从而提高学生学习的自主性与创造性。

> ➢ **适用对象**

本书可作为高职高专院校的计算机专业英语教材、IT 职业英语考试的参考教材或各种相关培训班的专业英语教材，也可供计算机工程技术人员或广大英语爱好者使用。

> ➢ **关于编者**

本书的 UNIT1、UNIT5 由王旭升编写，UNIT2、UNIT9 及附录由权小红编写，UNIT3 由赵香会编写，UNIT4 由刘穗编写，UNIT6 由贺萌编写，UNIT7 由易琼编写，UNIT8 由李琳编写，UNIT10 由权小红、李春华共同编写，权小红负责全书的统稿工作。

在本书的编写过程中，我们参阅了许多计算机英语及英语学习网站的参考资料，这里谨向资料作者致以衷心的感谢！由于编者水平有限，加之时间仓促，疏漏与不妥之处在所难免，敬请读者不吝赐教。

编 者
2012 年 5 月

# 目 录

## UNIT 1 .................................................................................................................. 1
- TOPICS ................................................................................................................ 1
- TEXT　The History of Computers in a Nutshell ........................................... 2
- EXERCISES ......................................................................................................... 5
- SUPPLEMENTARY　Famous People in the History of IT ............................ 8
- CONVERSATION　Etiquette and Tips to Successful Job Interview .......... 11
  - Example ......................................................................................................... 13
  - Practice .......................................................................................................... 14
  - Tips ................................................................................................................ 14
- WRITING　How to Write a Resume ............................................................. 16
  - Example ......................................................................................................... 16
  - Practice .......................................................................................................... 18
  - Tips ................................................................................................................ 18

## UNIT 2 ................................................................................................................ 21
- TOPICS .............................................................................................................. 21
- TEXT　Basic Components of a Modern Computer ..................................... 22
- EXERCISES ....................................................................................................... 24
- SUPPLEMENTARY　How to Buy a Computer ............................................ 28
- CONVERSATION　Business Etiquettes for Western Banquets ................. 30
  - Example ......................................................................................................... 32
  - Practice .......................................................................................................... 33
  - Tips ................................................................................................................ 33
- WRITING　How to Write a Perfect Professional E-mail in English ......... 35
  - Example ......................................................................................................... 36
  - Practice .......................................................................................................... 37
  - Tips ................................................................................................................ 38

## UNIT 3 ................................................................................................................ 40
- TOPICS .............................................................................................................. 40
- TEXT　What Is Computer Software? ............................................................ 41
- EXERCISES ....................................................................................................... 44
- SUPPLEMENTARY　9 Common Windows 7 Problems ............................. 47
- CONVERSATION　Brainstorming ................................................................ 50
  - Example ......................................................................................................... 51

Practice ........................................................................................................................... 52

　　Tips ................................................................................................................................. 53

　WRITING　How to Write Application and Recommendation Letters ................................. 54

　　Example .......................................................................................................................... 55

　　Practice ........................................................................................................................... 56

　　Tips ................................................................................................................................. 57

## UNIT 4 ................................................................................................................................... 60

　TOPICS ............................................................................................................................... 60

　TEXT　Software Engineering .............................................................................................. 61

　EXERCISES ........................................................................................................................ 65

　SUPPLEMENTARY　About 4G .......................................................................................... 67

　CONVERSATION　Telephone Marketing Skills ................................................................. 69

　　Example .......................................................................................................................... 71

　　Practice ........................................................................................................................... 72

　　Tips ................................................................................................................................. 73

　WRITING　How to Write IOU and Receipt in English ....................................................... 74

　　Example .......................................................................................................................... 74

　　Practice ........................................................................................................................... 75

　　Tips ................................................................................................................................. 76

## UNIT 5 ................................................................................................................................... 78

　TOPICS ............................................................................................................................... 78

　TEXT　An Introduction to SQL ........................................................................................... 79

　EXERCISES ........................................................................................................................ 82

　SUPPLEMENTARY　Database Management System and Database Model ..................... 84

　CONVERSATION　Booking Hotels .................................................................................... 89

　　Example .......................................................................................................................... 90

　　Practice ........................................................................................................................... 91

　　Tips ................................................................................................................................. 91

　WRITING　How to Write Business Contract ...................................................................... 93

　　Example .......................................................................................................................... 95

　　Practice ........................................................................................................................... 96

　　Tips ................................................................................................................................. 97

## UNIT 6 ................................................................................................................................... 99

　TOPICS ............................................................................................................................... 99

　TEXT　Computer Network and Internet ............................................................................ 100

　EXERCISES ...................................................................................................................... 103

SUPPLEMENTARY  Commonly Used Computer Network ............................................................. 105
CONVERSATION  Bargaining Skills ........................................................................................ 107
    Example ............................................................................................................................. 107
    Practice .............................................................................................................................. 108
    Tips .................................................................................................................................... 109
WRITING  How to Write a Lost and Found Notice ................................................................. 109
    Example ............................................................................................................................. 110
    Practice .............................................................................................................................. 111
    Tips .................................................................................................................................... 112

# UNIT 7 ................................................................................................................................ 115
TOPICS ...................................................................................................................................... 115
TEXT  E-commerce (electronic commerce or EC) ................................................................. 116
EXERCISES ............................................................................................................................... 118
SUPPLEMENTARY  About Taobao ....................................................................................... 121
CONVERSATION  How to Place an Order .......................................................................... 123
    Example ............................................................................................................................. 124
    Practice .............................................................................................................................. 125
    Tips .................................................................................................................................... 125
WRITING  How to Write Letter of Complaint and Letter of Apology ................................... 126
    Example ............................................................................................................................. 126
    Practice .............................................................................................................................. 128
    Tips .................................................................................................................................... 129

# UNIT 8 ................................................................................................................................ 132
TOPICS ...................................................................................................................................... 132
TEXT  Information Security ..................................................................................................... 133
EXERCISES ............................................................................................................................... 136
SUPPLEMENTARY  Viruses Can Eat Your Computer Alive ................................................. 138
CONVERSATION  Entertaining Clients ................................................................................ 140
    Example ............................................................................................................................. 141
    Practice .............................................................................................................................. 142
    Tips .................................................................................................................................... 142
WRITING  How to Write Effective Letters of Appreciation and Congratulation .................. 144
    Example ............................................................................................................................. 144
    Practice .............................................................................................................................. 145
    Tips .................................................................................................................................... 146

# UNIT 9 ................................................................................................................................ 148

TOPICS ........................................................................................................................... 148
TEXT　What Is Multimedia? ...................................................................................... 149
EXERCISES ................................................................................................................. 152
SUPPLEMENTARY　Where Is Virtual Reality? ........................................................ 154
CONVERSATION　8 Proverbs for Business Negotiation ......................................... 156
　　Example ............................................................................................................... 157
　　Practice ................................................................................................................ 158
　　Tips ...................................................................................................................... 158
WRITING　How to Write a Notice or an Inviation Letter .......................................... 159
　　Example ............................................................................................................... 160
　　Practice ................................................................................................................ 162
　　Tips ...................................................................................................................... 163

## UNIT 10 ............................................................................................................... 166
TOPICS ........................................................................................................................... 166
TEXT　Embedded System .......................................................................................... 167
EXERCISES ................................................................................................................. 169
SUPPLEMENTARY　Inception Review ...................................................................... 172
CONVERSATION　Understanding Party Culture in the West .................................. 174
　　Example ............................................................................................................... 176
　　Practice ................................................................................................................ 177
　　Tips ...................................................................................................................... 177
WRITING　How to Write the Abstracts of Scientific and Technological Theses ...... 178
　　Example ............................................................................................................... 180
　　Practice ................................................................................................................ 182
　　Tips ...................................................................................................................... 183

**附录　计算机英语常用词汇术语表** ................................................................................. 185

# UNIT 1

## TOPICS

- How people dealt with numbers and data in ancient times?
- How do you think W.W. II might have been different if the ENIAC had not been invented then?
- Who came up with the idea of using "binary code" to store programs for computers?
- What's your earliest computer memory?
- What is your vision of the computer technology in the next 20, 50, 100 years?
- Tips for job interview.
- How to write a resume?

# The History of Computers in a Nutshell

Computers and computer **applications** are on almost every aspect of our daily lives. As like many ordinary objects around us, we may need clearer understanding of what they are. You may ask "What is a computer?" or "What is a **software**", or "What is a **programming language**?" First, let's examine the history.

The term Computer, originally meant a person capable of performing **numerical** calculations with the help of a **mechanical** computing device. The evolution of computers started way back in the **era** before Christ. **Binary arithmetic** is at the **core** of computer systems. The history of computers starts out about 2000 years ago with the birth of the **abacus**, a wooden rack holding two horizontal wires with beads strung on them. The invention of **logarithm** by John Napier and the invention of slide rules by William Oughtred were significant events in the evolution of computers from these early computing devices.

If you look at the timeline of the evolution of computers, you will notice that the first computers used **vacuum tubes** for **circuitry** and **magnetic drums** for memory, and were often **enormous**, taking up entire rooms. They were very expensive to operate and in addition to using a great deal of electricity, generated a lot of heat, which was

application [ˌæpliˈkeiʃən] n. 申请；应用程序

software [ˈsɔftwɛə] n. 软件；软体；软设备

programming language  编程语言

numerical [njuːˈmerik] adj. 数字的；数值的

mechanical [miˈkænikəl] adj. 机械的；体力的；手工操作的

era [ˈiərə] n. 纪元，年代；历史时期，时代

binary [ˈbainəriː] adj. 双重的；二态的；二元的；二进制的 n. 二进制数；双子星

arithmetic [əˈriθmətik] n. 算术，计算；算法

core [kɔː] n. 中心，精髓；果核

abacus [ˈæbəkəs] n. 算盘

logarithm [ˈlɔːgə,riðəm] n. 对数

vacuum tubes  真空管

circuitry [ˈsəːkitri] n. 电路，线路

magnetic drum  磁鼓

enormous [iˈnɔːməs] adj. 巨大的，庞大的

often the cause of **malfunctions**.

The UNIVAC and ENIAC computers are examples of first-generation computing devices. The UNIVAC was the first commercial computers delivered to a business client, the U.S. Census Bureau in 1951. These computers were expensive and **bulky**. They used machine language for computing and could solve just one problem at a time. They did not support **multitasking**. Input was based on punched cards and paper tape, and output was displayed on printouts.

**Transistors** replaced vacuum tubes and **ushered** in the second generation of computers. The transistor was invented in 1947 but did not see widespread use in computers until the late 50s. The transistor was far **superior** to the vacuum tube, allowing computers to become smaller, faster, cheaper, more energy-efficient and more reliable than their first-generation **predecessors**. Though the transistor still generated a great deal of heat that subjected the computer to damage, it was a vast improvement over the vacuum tube. Second-generation computers still relied on punched cards for input and printouts for output.

Second-generation computers moved from **cryptic** binary machine language to **symbolic**, or **assembly**, languages, which allowed programmers to specify instructions in words. High-level programming languages were also being developed at this time, such as early versions of COBOL and FORTRAN. These were also the first computers that stored their instructions in their memory, which moved from a magnetic drum to magnetic core technology.

malfunction [mæl'fʌŋkʃən] n. 故障，功能障碍；失灵  vi. 失灵；发生故障

bulky ['bʌlki:] adj. 庞大的，笨重的，体积大的

multitasking[,mʌlti'tɑːskɪŋ] n. 多任务（处理）

transistor [træn'sɪstə(r)] n. 晶体管；晶体管收音机

usher ['ʌʃə(r)] n. 带位员；招待员
vt. 引导；引入

superior [sjuː'pɪəriə(r)] n. 上级；高手；上标  adj. 上层的；上好的；出众的

predecessors ['priːdɪsesə(r)] n. 祖先；前任；原有事物

cryptic ['krɪptɪk] adj. 隐秘的；秘密的；用密码的；隐晦的

symbolic [sim'bɔlik] adj. 象征的；符号的  n. 代号

assembly [ə'sembli] n. 集会；装配；[计] 汇编

The use of integrated circuits ushered in the third generation of computers. Their use increased the speed and efficiency of computers. **Operating systems** were the human **interface** to computing operations and keyboards and monitors became the input-output devices. COBOL, one of the earliest computer languages, was developed in 1959–1960. BASIC came out in 1964. It was designed by John George Kemeny and Thomas Eugene Kurtz. Douglas Engelbart invented the first mouse prototype in 1963. Computers used a video display terminal (VDT) in the early days. The invention of Color Graphics Adapter in 1981 and that of Enhanced Graphics Adapter in 1984, both by IBM added "color" to computer displays. All through the 1990s, computer monitors used the CRT technology. LCD replaced it in the 2000s. Computer keyboards evolved from the early typewriters. The development of computer storage devices started with the invention of Floppy disks, by IBM again.

operating systems 操作系统
interface ['ɪntəfeɪs] n. 界面；接口

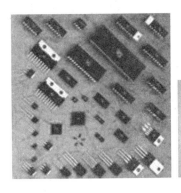

Thousands of **integrated circuits** placed onto a **silicon** chip made up a **microprocessor**. Introduction of microprocessors was the **hallmark** of fourth generation computers.

● Intel produced large-scale integration circuits in 1971. Microprocessors came up during the 1970s. Ted Hoff, working for Intel introduced 4-bit 4004.

● In 1972, Intel introduced the 8080

integrated circuit 集成电路
silicon ['sɪlɪkən] n. 硅
microprocessor [ˌmaɪkrəʊ'prəʊsesə(r)] n. 微处理器
hallmark ['hɔːlmɑːk] n. 纯度标记；标志；特征 vt. 标纯度

microprocessors.

● In 1974, Xerox came up with Alto **workstation** at PARC. It consisted of a monitor, a graphical interface, a mouse, and an Ethernet card for networking.

● Apple Computers brought out the Macintosh personal computer on January 24 1984.

● By 1988, more than 45 million computers were in use in the United States. The number went up to a billion by 2002.

workstation[wəːksˈteiʃən] n. 工作站

The fifth generation computers are in their development phase. They would be capable of massive parallel processing, support voice recognition and understand natural language. The current advancements in computer technology are likely to transform computing machines into intelligent ones that possess self organizing skills. The evolution of computers will continue, perhaps till the day their processing powers equal human intelligence.

 EXERCISES

Ⅰ. Match the terms and the interpretations.

1. Abacus        (a) A low-level programming language for computers, microprocessors, microcontrollers, and other programmable devices in which each statement corresponds to a single machine language instruction. It is specific to a certain computer architecture, in contrast to most high-level programming languages, which may

be more portable.

2. Multitasking  (b) A manual aid to calculating that consists of beads or disks that can be moved up and down on a series of sticks or strings within a usually wooden frame.

3. Workstation  (c) A method where multiple tasks, also known as processes, are performed during the same period of time.

4. Assembly Language  (d) It is a multipurpose, programmable device that accepts digital data as input, processes it according to instructions stored in its memory, and provides results as output.

5. Microprocessor  (e) A high-end microcomputer designed for technical or scientific applications. Intended primarily to be used by one person at a time, they are commonly connected to a local area network and run multi-user operating systems. The term has also been used to refer to a mainframe computer terminal or a PC connected to a network.

II. Are the following statements True (T) or False (F)?

1. (    ) 0 and 1 are the binary numbers.
2. (    ) Abacus is considered to be the first calculator.
3. (    ) ASCII is the abbreviation for American Standard Code for Information Interchange.
4. (    ) Because of transistors, computers have given off no heat.
5. (    ) Vacuum tubes were replaced by integrated circuits.
6. (    ) First generation computers used vacuum tubes for circuitry and floppy disk for memory.
7. (    ) Second-generation computers still relied on punched cards for input and printouts for output.
8. (    ) Assembly languages came into play in the third generation computers.
9. (    ) Douglas Engelbart invented the first mouse prototype in 1963.
10. (    ) Fifth generation computers would be capable of massive parallel processing, support voice recognition and understand natural language.

III. Translate the following words and phrases into Chinese.

1. Binary system    _____

2. CRT _____
3. LCD _____
4. Integrated circuits _____
5. Massive parallel processing _____
6. Voice recognition _____
7. Operating system _____
8. Interface _____
9. Floppy disks _____
10. Color Graphics Adapter _____

IV. Translate the following Chinese statements into English.

1. 公元前 5 世纪，中国人发明了算盘，广泛应用于商业贸易中。算盘被认为是最早的计算设备，并一直使用至今。

2. 1943 年到 1959 年的计算机通常被称做第一代计算机。此时期的计算机使用真空管，所有的程序都是用机器码编写的，使用穿孔卡片。

3. 1964 年到 1972 年的计算机一般被称为第三代计算机。此时期的计算机大量使用集成电路，典型的机型是 IBM360 系列。

4. 与整个人类的发展历程及传统科学技术相比，计算机的历史才刚刚开始。

5. 计算机是由硬件系统和软件系统组成的，硬件包括中央处理器、存储器、输入设备和输出设备，软件系统包括系统软件和应用软件。

V. Fill in each of the blanks with one of the following words.

upon   led up to   for   analyze   capable of
too   mechanical   calculate   solved   instead of

Since civilizations began, many of the advances made by science and technology have depended _____ the ability to process large amounts of data and perform complex mathematical calculations. For thousands of years, mathematicians, scientists and businessmen have searched _____ computing machines that could perform calculations and _____ data quickly and efficiently. One such device was the abacus.

The abacus was an important counting machine in ancient Babylon, China, and throughout Europe where it was used until the late middle ages. It was followed by a series of improvements in _____ counting machines that _____ the development of accurate mechanical adding machines in the 1930's. These machines used a complicated assortment of gears and levers to perform the calculations but they were far _____ slow to be of much use to scientists. Also, a machine capable of making simple decisions such as which number is larger was needed. A machine _____ making decisions is called a computer.

In June 1943, work began on the world's first electronic computer. It was built at the University of Pennsylvania as a secret military project during World War II and was to be used to _____ the trajectory(弹道，轨迹) of artillery shells. It covered 1500 square feet and weighed 30 tons. The project was not completed until 1946 but the effort was not wasted. In one of its first demonstrations, the computer _____ a problem in 20 seconds that took a team of mathematicians three days. This machine was a vast improvement over the mechanical calculating machines of the past because it used vacuum tubes _____ relay switches. It contained over 17,000 of these tubes, which were the same type tubes used in radios at that time.

The invention of the transistor made smaller and less expensive computers possible. Although computers shrank in size, they were still huge by today's standards. Another innovation to computers in the 1960's was storing data on tape instead of punch cards. This gave computers the ability to store and retrieve data quickly and reliably.

SUPPLEMENTARY

## Famous People in the History of IT

Almost everyone uses computers these days for everything from shopping to working to playing games. But have you ever stopped to think about where all this amazing technology came from?

Who invented it all? Well, behind every company, programming language or piece of software, there is a person—or sometimes a team of people—who turned ideas into reality. We've all heard of Bill Gates, the founder of Microsoft and one of the richest men in history. Equally famous is Steve Jobs, the person who, along with Steve Wozniak, started Apple computers. However, there are hundreds of other people, from early pioneers to later geniuses, who aren't as well known but who deserve recognition for the work they did in advancing the world of computing.

Bill Gates

Charles Babbage

One of the first people to conceive of computers was Charles Babbage, an English mathematician and analytical philosopher who drew up plans for the first programmable computer called the Difference Engine. George Boole came up with a way of describing logical relations using mathematical symbols—now called Boolean logic—that is the basis of all modern computer processes. Vannevar Bush first proposed an idea in 1945 he called "memex", which we now know as "hypertext". Another notable figure in early computing was Alan Mathison Turing, an Englishman known as the "father of computer science". He invented the Turing Test, which is a way to find out if a computer is acting like a machine or a human. Another English computer scientist, Edgar Frank Codd, is known for inventing the "relational" model for databases, a model which is still in use today.

Mathison Turing

As computing became more complicated, people needed a way to make it easier to tell computers what to do—in other words, they needed ways to program the computers. These computer instruction systems became known as computer, or programming, languages. FORTRAN, the first widely used high-level programming language, was invented by an American computer scientist, John Warner Backus. Other notable North American inventors of programming languages include Dennis Ritchie, author of the C programming language, Larry Wall, creator of Perl, and Canadian James Gosling, known as the father of Java. Two men from Denmark are responsible for writing two other famous programming languages. Bjarne Stroustrup came up with C++ and Rasmus Lerdorf devised PHP. Dutchman Guido van Rossum wrote the Python programming language, while the Japanese computer scientist, Yukihiro Matsumoto, made a language called Ruby.

One of the uses of programming languages is to create operating systems, which are essentially sets of instructions that allow computers to function. The most widely-used operating system in the world is Microsoft Windows, but there are other powerful ones that exist, such as Unix, created by Ken Thompson and his team at AT&T in 1969, and Linux, written by Linus Torvalds in 1991.

Linus Torvalds

Microsoft, of course, is the largest software company in the world, but there is another company, Intel, that is equally important when it comes to hardware. Intel was started by several people who are now legends in the computer world, including Robert Noyce and Gordon Moore. Moore is also famous for coming up with Moore's Law, which predicts the rapid increase of computer technology over time. Intel expanded rapidly during the 1980s and 1990s when a man named Andy Grove was in charge of the company.

Gordon Moore

Other notable figures in the evolution of the computer industry are Ralph Baer, inventor of the first home video game console, Seymour Cray, for many years the manufacturer of the world's fastest supercomputers, Richard Stallman, founder of the free software movement called GNU, and Tim Berners-Lee, the man who created the basis for the World Wide Web.

Through their creativity and hard work, all of these people contributed to shaping what we now experience as Information and Computer Technology. Every time you boot up a computer, play a video game or surf the Internet, try to remember the individuals who made these wonders possible.

# CONVERSATION

# Etiquette and Tips to Successful Job Interview

# 求职面试成功必备的礼仪和技巧

求职面试时，首先要明确面试前的三要素——When(时间)、Where(地点)、Who(联系人)。

**1. 基本面试礼仪**

(1) 约好面试时间后，一定要提前5～10分钟到达面试地点，以表示您的诚意，给对方以信任感，同时也可调整自己的心态，做一些简单的仪表准备，以免仓促上阵，手忙脚乱。

(2) 进入面试场合时不要紧张。如果门关着，应先敲门，得到允许后再进去。见面时要主动打招呼问好致意，称呼应当得体。在用人单位没有请您坐下时，切勿急于落座。离去时应询问"还有什么要问的吗"，得到允许后应微笑起立，道谢并说"再见"。

(3) 对用人单位提出的问题要逐一回答。对方给您介绍情况时，要认真聆听，可以在适当的时候点头或适当提问、答话。一般情况下不要打断用人单位的问话或抢问抢答，否则会给人急躁、鲁莽、不礼貌的印象。问话完毕，听不懂时可要求重复。当不能回答某一问题时，应如实告诉用人单位，含糊其辞和夸夸其谈会导致面试失败。

(4) 在整个面试过程中，应保持举止文雅大方，谈吐谦虚谨慎，态度积极热情。如果用人单位有两位以上主试人，回答谁的问题，您的目光就应注视谁，并应适时地环顾其他主试人以表示您对他们的尊重。谈话时，眼睛要适时地注视对方，不要东张西望，也不要低着头，激动地与用人单位争辩某个问题也是不明智的举动。

**2. 求职面试技巧之语言运用**

面试场上您的语言表达艺术将反映您的成熟程度和综合素养。对求职应试者来说，掌握语言表达的技巧无疑是重要的。那么，面试中怎样恰当地运用谈话的技巧呢？

(1) 口齿清晰，语言流利，文雅大方。

交谈时要注意发音准确，吐字清晰。还要注意控制说话的速度，以免磕磕绊绊，影响语言的流畅。为了增添语言的魅力，应注意修辞美妙，忌用口头禅，更不能有不文明的语言。

(2) 语气平和，语调恰当，音量适中。

面试时要注意语言、语调、语气的正确运用。打招呼时宜用上语调，加重语气并带拖音，以引起对方的注意。自我介绍时，最好多用平缓的陈述语气，不宜使用感叹语气或祈使句。声音的大小要根据面试现场情况而定。两人面谈且距离较近时声音不宜过大；群体面试且场地开阔时声音不宜过小，以每个用人单位都能听清您的讲话为原则。

(3) 语言要含蓄、机智、幽默。

讲话时除了表达清晰以外，适当的时候可以插进幽默的语言，使谈话增加轻松愉快的气氛，也会展示自己的良好气质和从容风度。尤其是当遇到难以回答的问题时，机智、幽默的语言会显示自己的聪明智慧，有助于化险为夷，并给人以良好的印象。

(4) 注意听者的反应。

求职面试不同于演讲，而是更接近于一般的交谈。交谈中，应随时注意听者的反应。根据对方的反应，要适时地调整自己的语言、语调、语气、音量、修辞，包括陈述内容。这样才能取得良好的面试效果。

### 3. 求职面试技巧之回答问题

(1) 把握重点，简捷明了，条理清楚，有理有据。

一般情况下回答问题要结论在先，议论在后，先将自己的中心意思表达清晰，然后再做叙述和论证。否则，如果长篇大论，会让人不得要领。面试时间有限，神经有些紧张，多余的话太多，容易走题，反倒会将主题冲淡或漏掉。

(2) 讲清原委，避免抽象。

用人单位提问总是想了解一些应试者的具体情况，切不可简单地仅以"是"和"否"作答。应针对所提问题的不同，有的需要解释原因，有的需要说明程度。不讲原委、过于抽象的回答，往往不会给主试者留下具体的印象。

(3) 确认提问内容，切忌答非所问。

面试中，如果对用人单位提出的问题一时摸不着头脑，以致不知从何答起或难以理解对方问题的含义，可将问题复述一遍，并先谈自己对这一问题的理解，请教对方以确认内容。对不太明确的问题，一定要搞清楚，这样才会有的放矢，不致答非所问。

(4) 有个人见解，有个人特色。

用人单位有时要接待若干名应试者，相同的问题会问若干遍，类似的回答也要听若干遍。因此，用人单位会有乏味、枯燥之感。只有具有独到的个人见解和个人特色的回答，才会引起用人单位的兴趣和注意。

(5) 知之为知之，不知为不知。

面试遇到自己不知、不懂、不会的问题时，回避闪烁、默不作声、牵强附会、不懂装懂的做法均不足取，诚恳坦率地承认自己的不足之处，反倒会赢得主试者的信任和好感。

# Example

## ⊠ Dialogue 1

**Interviewer:** Can you briefly explain what role you played in Customer Relationship Management process for Sales Force?

**Interviewee:** Yes, sure. I was a team leader while designing the Customer Relationship Management software for Sales Force. The project was regarding an international retail chain. With a team of 15 members I was able to successfully pin-point and utilize the factors affecting the footfall at the store and thus, helped the client achieve a growth of 30% in sales.

**Interviewer:** Great! So, how do you thing you fit in our organization?

**Interviewee:** I wish to make a career in IT industry and keeping in mind my long term goals, I believe that associating with this organization will be the most prudent step. In here, I would surely be able to further utilize my skills in the most effective manner and also, obtain a chance of acquiring newer skills and honing the same.

## ⊠ Dialogue 2

A: Good morning, my name is Ms Martin. You've applied for the Laboratory Assistant's position right?
B: Yes Ms Martin. I have.
A: Can you tell me why you replied to our advertisement?
B: Well, I've always enjoyed science and felt that this position would offer me an opportunity to extend my skills in this area.
A: Do you know exactly what you would be doing as a Laboratory Assistant?
B: A Laboratory Assistant helps to maintain scientific equipment, keeping a check on the supplies in the store, and preparing the chemicals for experiments.
A: What sort of student do you regard yourself as . . . did you enjoy studying while you were at school?
B: I suppose I'm a reasonable student. I passed all my tests and enjoyed studying subjects that interested me.
A: Now, do you have any questions you'd like to ask me about the position?
B: Yes. Ms Martin, could you tell me what hours I'd have to work, and for whom I'd be working?

## Practice

Attending job interviews can be intimidating! Being prepared will help you through the experience by increasing confidence, ensuring that you are presenting yourself appropriately and may ultimately bring success in securing the position. Here are some questions you might encounter in the job interview, practice job interview Q&A with your partner.

- Tell us about yourself?
- Why have you applied for this position?
- What is your understanding of the company?
- What are your strengths and weaknesses?
- What do you see yourself in 5 years?

## Tips

### ⊠ 工作面试常见30个问题

1. Tell me about yourself. 简要介绍一下您自己。
2. Why are you interested in this position? 您为什么对这份工作感兴趣?
3. What are your strengths? 您的优势是什么?
4. What is your biggest weakness? 您最大的弱点是什么?
5. Why do you feel you are right for this position? 为什么您认为自己适合这个职位?
6. Can you give me the highlights of your resume? 您的简历上有些什么值得特别关注的?
7. Why did you choose your major? 您为什么选择这个专业?
8. What are your interests? 您有哪些兴趣爱好?
9. What are your short and long term goals? 您的短期和长期目标是什么?
10. Tell me how your friends/family would describe you? 如果我向您的朋友或者家人询问对您的评价,您认为他们会怎样说?
11. Using single words, tell me your three greatest strengths and one weakness. 用简单的词语,描述您的三项最突出的优点和一个缺点。
12. What motivates you to succeed? 您争取成功的动力是什么?

13. What qualities do you feel are important to be successful? 哪些品质在您看来对成功是最重要的?

14. What previous experience has helped you develop these qualities? 哪些之前的经历帮助您获得了这些品质?

15. Can you give me an example of teamwork and leadership? 您能向我列举一个团队活动和领导力的例子吗?

16. What was your greatest challenge and how did you overcome it? 您经历过的最大的挑战是什么? 您是如何跨越它的?

17. Why should I hire you over the other candidates I am interviewing? 我为什么要从这么多应聘者中选择您呢?

18. Do you have any questions? 您有一些什么问题吗?

19. What are your compensation expectations? 您对于报酬有什么样的期望?

20. What was your greatest accomplishment in past time? 在过去的日子里,您觉得自己最大的成就是什么?

21. Have you ever been asked to do something unethical? If yes, how did you handle it? 曾经有人要求您去做一些不道德的事情吗? 如果有,您是怎么处理的呢?

22. What do you do if you totally disagree with a request made by your manager? 如果您完全不同意您上司的某个要求,您会怎么处理?

23. When in a group setting, what is your typical role? 您在团队中的作用通常是什么?

24. How do you motivate a team to succeed? 您怎么激励团队达到成功?

25. Have you been in team situations where not everyone carried their fair share of the workload? If so, how did you handle the situation? 如果您所处的团队中并不是每个成员都承担着相同的工作量,您将怎样处理这种情况?

26. How do you prioritize when you are given too many tasks to accomplish? 您怎样在一堆根本做不完的工作任务中区分轻重缓急?

27. Why are manhole covers round? 为什么下水道的井盖是圆的?

28. Tell me about a goal you set for yourself and how you accomplished it. 请和我讲一讲您为自己设定的某个目标以及是如何实现它的。

29. Do you typically achieve what you set out to do? 您一般情况下都能达到自己的既定目标吗?

30. What demotivates or discourages you? 有哪些因素可能会让您失去动力或信心?

# How to Write a Resume

## 如何写英文简历

对许多人来说,一份简约明快的英文简历是进入外企的"敲门砖"。那么,如何写好英文简历呢?

首先,语言简练。对于求职者来讲,目的明确、语言简练是其简历行之有效的基础。如在教育背景中写相关课程,不要为了拼凑篇幅,把所有的课程一股脑儿地都写上,如体育等。这样不太有效,别人也没耐心看。

其次,个人资料部分(PERSONALDATA),包括求职者的姓名、性别、出生年月等,与中文简历大体一致。第二部分为教育背景(EDUCATION),必须注意的是在英文简历中,求职者受教育的时间排列顺序与中文简历中的时间排列顺序正好相反,也即是从求职者的最高教育层次(学历)写起,至于低至何时,则无具体规定,可根据个人实际情况安排。

第三,社会工作。在时间排列顺序上亦遵循由后至前这一规则,即从当前的工作岗位写起,直至求职者的第一个工作岗位为止。求职者要将所服务单位的名称及自身的职位和技能写清楚。把社会工作细节放在工作经历中,这样会填补工作经验少的缺陷。例如,您在做团支书、学生会主席等社会工作时组织过什么活动,联系过什么事,参与过什么都可以一一罗列。而作为大学生,雇主通常并不指望您在假期工作期间会有什么惊天动地的成就。当然如果您有就更好了。

第四,所获奖励和作品(PRICE & PUBICATION)。将自己所获奖项及所发表过的作品列举一二,可以从另一方面证实自己的工作能力和取得的成绩。

另外,大多数外企对英语(或其他语种)及计算机水平都有一定的要求,个人的语言水平、程度可在此单列说明。

## Example

### Resume

**Jane R. Doe**
910 Oak Street, Verona, CA12111 555-550-111-1111
abcd @ yahoo.com
**EXPERIENCE**

**Computer Company, Software Engineer**

August 2003-present

Software Engineer on Company Soft Manager. Duties include developing current release using C++ and Java, assisting in design of next release (J2EE), traveling to standard meetings at SNIA SMIS-S to represent Company Soft Manager, traveling to SNW, a semi-annual consumer conference to showcase product, and working closely with new developers in India Tech Center.

**Computer Company Training Program**

June 2003-August 2003

Member of the Computer Company Bootcamp program, an intensive 3 month training program for choice software engineers. The three month program covered advanced topics in software engineering, SQL, C++, J2EE, XML, Windows 2000 Server, Unix, UML, and various Company products.

**Consultant**

January 2003-June 2003

Consultant for high school in the outer Boston area. Tutored the programming instructors for the AP programming class in the Java programming language. Helped to set up development environment for the classroom.

**Company Inc, Software Engineer CO-OP**

January 2001-September 2001

Developer on the Company engine team. Worked on new functionality in the 7.0 release of the Company Dynamic Sourcing Engine. Developed in C++ in Unix and Windows Visual Studio. Also worked on a solo project to add multithreaded capabilities to Company's engines.

**TECHNICAL**

Languages: C++, Java, C, ASP.NET, SQL

Applications: MS Visual Studio, Eclipse

Application Server: JBoss, Tomcat

Operating Systems: Windows, Unix, Linux

Database Systems: SQL Server, MySQL

Certifications: CCNA, Unicenter Certified Engineer

**EDUCATION**

ABC College, Troy, NY, May 2002

Major: Computer Science, Minor: Management

**ACTIVITIES**

Brother of Delta Chapter of Delta Chi Fraternity

-President (January 2001-May 2002)

-Scholarship winner at Delta Chi Leadership Conference

# Practice

根据如下的企业招聘条件,写一份个人简历。

**Position: Software Developer (IT support department)**

**Responsibilities:**
(1) Specification of software projects.
(2) Setup and build the software development standard, guideline, and quality assurance, risk control for enterprise software development.
(3) Managing day-to-day Software development related activities, ensuring that all projects are completed in a thorough and timely fashion.
(4) Reviewing all technical materials for completeness and accuracy, thorough documentation.

**Requirements:**
(1) Computer Science or Engineering degree.
(2) Expertise in NET framework, SQL database design, web development and GUI-based tool development.
(3) Expertise in C#, PHP programming.
(4) Knowledge of data mining is advantages.
(5) Knowledge of CMMI and other quality methodology is advantages.
(6) Hard-working, self-motivated person with good attention to detail and a good team spirit.

# Tips

## ✉ 面试之后的感谢信

　　面试结束后,大部分求职者都会觉得面试已经结束,等着公司的通知就行了。其实,面试后还有许多工作要做,写面试后的感谢信就是必须要做的工作。面试感谢信是对面试官的尊重与礼貌,可以加深面试官对求职者的良好印象。面试后的感谢信写得好,会增加求职者的录用机会。

面试之后的感谢信之一

Dear Mr./Ms. [hiring manager last name]

It was a pleasure meeting with you yesterday, and I thank you for your time. I appreciate the fact that you've taken the time to acquaint me with the team, discussing about the [position] and presenting the company background.

After meeting with you and further observing the company's operations, I am convinced that my professional experience and skills coincide well with the position needs.

Now that I have met you and know more about the job requirements, I am even more excited about the opportunity of working as a member of your team.

Having the motivation to exceed prior expectations, as I briefed during the interview, I am ready to handle the challenges that you offer me and would definitely be a value added addition to the team and to the company.

I remain confident that my competencies are a good match for your needs, and hope to be among those in consideration for the job.

If there are any further questions you would like to ask me, please contact me via E-mail or telephone. Of course, I will be available for future interviews as needed.

I will look forward to speaking with you again soon.

Your Sincerely,
[Your signature]
[Name]

## 面试之后的感谢信之二

Dear Mr./Ms. [hiring manager last name]

I would like to thank you for interviewing me today for the marketing analyst position at IBM. I left the interview with a renewed esteem for the company and trust that you've recognized my interest in being part of your team.

As you presented the job, it offers the professional challenges and future growth that I always want for my career.

I am positive that I can contribute my enthusiasm and dedication to the position and convinced that my background and skills equip me more than adequately for the job requirement.

In addition to my enthusiasm and to reiterate my point further, I will add to the position—my strong verbal/written communication skills and the ability to work under pressure/workload and timelines.

My marketing background will help me learn quickly the job needs and adjust/adapt myself to the company culture.

Please do not hesitate to contact me, if you need more information about my background and qualifications. If you have any further questions, I can make myself available for any further discussions and interviews.

Again, I appreciate the time you took to interview me.

I am very interested in working for you and look forward to hearing from you soon about the position.

Your Sincerely,
[Your signature]
[Name]

## 面试之后的感谢信之三

Dear …

Thank you for taking the time to interview me for the [job title, example—software programmer position] today.

I am grateful for the way you presented me the job [in details… if it was the second interview] and the company's work culture. I'd like to tell you that I am impressed with the company's reputation as well as the career growth/opportunity that you offer.

As I am very much interested in this position, my hope is that my competencies fit well with your requirements.

I am eager to bring my knowledge to the position, and believe that my [A,B,C] extensive experience I've already developed make me a good candidate.

I look forward to provide more information about my qualifications and the possibility of working with you.

Thanks you again for your time and consideration.

Your Sincerely,
[Your Name]

# UNIT 2

## TOPICS

- What are the basic components of a modern computer?
- What is the role of CPU in a PC?
- How does the CPU work?
- What is dual core CPU?
- What is the main job of the motherboards in a PC?
- What are the types of motherboards?
- Why could we say the hard disk is the "data center" of the PC?
- What is RAM?
- Situational conversations on dining.
- How to write a perfect professional E-mail in English?

# Basic Components of a Modern Computer

Computers **are composed of** two basic specifications, hardware and software. Hardware consists of the actual physical parts of the machine. Hardware can **pertain to** many different things including: the CPU, memory, hard drive, mouse, keyboard, motherboard, etc. So hardware refers to a physical piece of a computer. The key idea is that the item is something you can touch. This compares to software which is not **tangible** in any way. You can't pick it up or weigh it.

## CPU—Central Processing Unit

Think of the CPU as the "Brain" of a computer and of all your hardware. It is indeed responsible for **interpreting** and executing most of the commands from the computer's hardware and software. Its job is deceptively simple—to execute programs. Keep in mind that a program is simply a sequence of stored instructions. However, the process of retrieving data, **decoding** it, executing it, and storing the results, is far from simple. It involves many different parts of the CPU; each specialized to handle a specific job. On top of that, these simple steps are broken into many smaller operations and have been extensively modified and added to in order to improve performance.

## Motherboard or Mainboard

The motherboard and the CPU are the most important hardware components. The motherboard contains **slots** for the CPU, memory, and I/O devices. The core of each mainboard is a pair of chips collectively referred to as "the Chip Set". They sit in the middle of the mainboard and are connected to everything else. In current designs,

be composed of 由……组成

pertain to 属于，关于，附属，适合，相称

tangible ['tændʒəbəl] adj. 可触摸的，实际的，有形的，确凿的

interpreting [in'tə:pritiŋ] n. 口译，解释；动词 interpret 的现在分词

decode ['di:'kəud] v. 解码，译解，译码

slot [slɔt] n. 狭缝，槽，投币口，时间段；职位 v. 插入，留细长的孔，放置

one chip called the Northbridge sits between and connects the three high speed devices: CPU, memory, and PCI Express or AGP video port. It is then connected to a second chip called the Southbridge that provides logic for all the slow speed devices: the keyboard, mouse, modem port, printer port, IDE controller, PCI, USB, and any other devices. The motherboard is attached to a tray in the bottom or side of the case by nine screws that **screw** into metal "**standoffs**" that keep the bottom of the mainboard a safe distance from the metal of the case. Everything else plugs into the mainboard.

screw [skru:] n. 螺旋桨，螺旋，螺钉，螺丝，螺状物 v. 拧紧，拧

standoff ['stænd.ɔ:f] adj. 冷淡的，有支架的 n. 离岸；避开，冷淡，抵消，和局，平衡

## Hard Disk

The hard disk is also a very important piece of hardware. The hard disk drive in your system is the "data center" of the PC. It is here that all of your programs and data are stored between the **occasions** that you use the computer. Your hard disk are the most important of the various types of permanent storage used in PCs (the others being floppy disks and other storage media such as CD-ROMs, tapes, removable drives, etc.) The hard disk differs from the others primarily in three ways: size (usually larger), speed (usually faster) and permanence (usually fixed in the PC and not removable).

occasion [ə'keiʒən] n. 场合，时机，理由，机会，盛大场面 vt. 引起，致使

## RAM—Random Access Memory

Computer memory is called RAM. Certain motherboards only work with certain types of RAM. Because RAM is temporary memory, or random access memory, it is not used in the same way your hard drive is used. When you open a new application or window the data is transferred by CPU into the RAM. Your computer then runs the program from RAM so that you can easily access files and folders.

If a program or file is too large to fit in RAM, the overflow is stored in virtual memory. In Windows, virtual memory is called the paging file or swap. The less RAM that your computer has the more paging is used, which ultimately causes a system slow down. More RAM will give you better performance and faster speed.

**Keyboard and Mouse**

Users like you will need a device such as a keyboard to enter your words and commands for the computer. Each time you type a key on the keyboard, it sends a signal through the circuit **underneath** called **Key Matrix Circuit** to the computer. The connection from a keyboard to the computer is accomplished with a cable to a PS/2 or USB connection.

underneath [ˌʌndəˈniːθ] adv. 在下面 n. 下部, 底部  prep. 在……下面
Key Matrix Circuit  键盘矩阵电路

Another major component is the mouse, which is mostly used for clicking a button, checkbox, option box, or to navigate around the screen on the monitor. There are basically ball mouse, optical mouse, and laser mouse types. A ball mouse is a mouse with a round ball under the casing, moving the mouse on a mouse pad will rotate the ball and then send the signal to the computer for the **navigation**. An optical or laser mouse both use optical to send signal to the computer, the difference is an optical mouse uses light-emitting **diode** or photo diode to detect the movement, while laser mouse uses **infrared laser** diode to sense the motions.

navigation [nævigˈeiʃn; nævəˈgeiʃən] n. 航海, 航空, 导航, 领航, 航行
diode [ˈdaiəud] n. 二极管
infrared laser  红外线激光器

# EXERCISES

Ⅰ. Match the terms and the interpretations.

1. Power Supply Unit (PSU)  (a) A specialized circuit designed to rapidly manipulate and alter memory in such a way so as to accelerate the building of images in a frame buffer intended for output to a display.

2. Multi-Core Processor  (b) A specification to establish communication between devices and a host controller(usually a personal computer), which has effectively replaced a variety of earlier interfaces such as serial and parallel ports.

3. Universal Serial Bus (USB)  (c) A single computing component with two or more independent actual processors (called "cores"), which are the units that read and execute program instructions.

4. Graphics Processing Unit (GPU)  (d) The rows and columns are made up of wiring. Depressing a key connects a circuit in the matrix, which causes the tone generation mechanism to be triggered.

5. Key Matrix Circuit  (e) It powers all the computer's main components.

II. Are the following statements True (T) or False (F)?

1. (   ) With the development of computer, the physical size of the CPU has often become bigger and bigger.
2. (   ) Portable computers can fit in a briefcase or even in the palm of your hand.
3. (   ) The purpose of a cache is to enable the CPU to access recently used information very quickly.
4. (   ) The CPU comprises the control unit and the memory.
5. (   ) The movement of electronic signals between main memory and the ALU as well as the control signal between the CPU and input/output devices are controlled by the control unit of the CPU.
6. (   ) The control unit performs all the arithmetic and logical (comparison) functions.
7. (   ) The Central Processing Unit (CPU) is the heart of the computer systems.
8. (   ) A motherboard is the electronic highways for the electronic currents to obtain from a module to another module.
9. (   ) The integrated style of motherboard includes many components on the board itself.
10. (   ) The purpose of output hardware is to collect data and convert it into a form suitable for computer processing.

III. Translate the following words and phrases into Chinese.

1. Chip Set                    _____
2. Key Matrix                  _____
3. Portable Computer           _____
4. Multi-Core Processor        _____
5. ALU                         _____
6. PAD                         _____

7. UPS　　　　＿＿＿＿＿＿＿＿＿＿＿＿＿＿＿＿＿

8. BIOS　　　＿＿＿＿＿＿＿＿＿＿＿＿＿＿＿＿＿

9. CMOS　　＿＿＿＿＿＿＿＿＿＿＿＿＿＿＿＿＿

10. HUB　　　＿＿＿＿＿＿＿＿＿＿＿＿＿＿＿＿＿

Ⅳ. Translate the following Chinese statements into English.

1. 输入硬件的用途是收集数据，并将其转换成适合于计算机加工处理的格式。

_____

2. 如果计算机有一个设计优良的CPU，那么你就能在极短的时间内，执行高度复杂的任务。

_____

3. 手册上说这个播放器需要32位的声卡。

_____

4. 计算机的主存容量越大，你能完成的任务就越多。

_____

5. 我的鼠标好像不怎么好用。

_____

Ⅴ. Can you recognize them? Write down their names.

Figure 1 (　　　　　　　)

Figure 2 (　　　　　　　)

Figure 3 (　　　　　　　)

Figure 4 (　　　　　　　)

Figure 5 (                    )

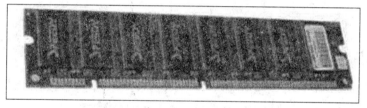

Figure 6 (                    )

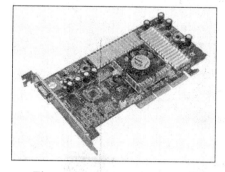

Figure 7 (                    )

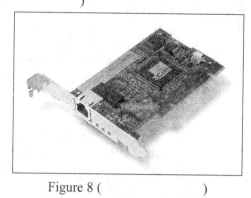

Figure 8 (                    )

Figure 9 (                    )

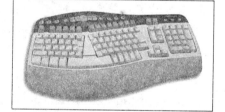

Figure 10 (                    )

Figure 11 (           )

Figure 12 (           )

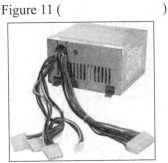
Figure 13 (           )

Figure 14 (           )

 SUPPLEMENTARY

## How to Buy a Computer

Buying a computer means investigating many features: RAM(Random-Access Memory), processor speed, graphics capability, hard disk space and so on. Here's how to start.

## Things you'll need

- **Hardware:**
CPU(Computer Processing Unit)  CD-ROM Drives  CD-RW  Burners...
DVD-ROM Drive  RAM  Computer Keyboards  Computer Mice
Computer Monitors  Computer Speakers  Printers  Hard Disks  Modems
Floppy Disk Drive  Sound Cards ...
- **Software:**
Microsoft Windows  Microsoft Office  Anti-virus Software ...

## Instructions

1. Choose a specialty store, consumer electronics store, retail chain, limited service discount chain, local computer builder or mail-order/Internet vendor, based on your hardware and service needs.

2. Buy when you need to buy. No matter how long you wait for the best deal, the same configuration will cost less in six months.

3. Decide which features you'll need based on what you're going to do with the computer. For example, if you're going to be creating graphics, sound and video, you'll want plenty of RAM. If you're going to be doing heavy computational tasks (searching large databases, watching video), you'll want a super processor.

4. Decide if you want a laptop, which you can carry around with you, or a desktop model.

5. Choose a computer brand based on quality, price and technical support.

6. Based on your likely needs, determined earlier, figure out the core configuration you need, including processor and speed, amount of RAM and hard drive size.

7. Determine additional drives you need: CD-ROM, DVD-ROM, CD recorder, Zip.

8. Select peripherals and additional hardware such as modems, sound cards, video cards and speakers.

9. Decide how many extra internal card slots and disk-drive bays you'll need in order to allow room for future expansion.

10. Determine what pre-installed software you want or need. Get at least an operating system, such as Windows, an anti-virus program, and programs for word processing, spreadsheets, databases and keeping your checkbook.

11. Choose the length of warranty or service coverage appropriate to your needs.

## Tips & Warnings

1. Keep abreast of the latest technology by reading the new-product reviews in magazines and on the Web. Find out what hardware and soft-ware are included with the models you're considering, and use that as a basis for comparing prices. Ask "What's the catch?" if a price seems too low.
2. Understand that RAM is where your computer temporarily stores data to be processed. Although more RAM is better, you don't need much if you restrict your computer use to simple tasks (word processing, check balancing).
3. Games, however, can require lots of RAM; graphics and sound are other space hogs.
4. Today's processors are usually fast enough for all but the most demanding applications, such as streaming video.

CONVERSATION

## Business Etiquettes for Western Banquets

## 工作场合西餐礼仪

1. Electronic devices.

Turn off or silence all electronic devices before entering the restaurant. If you forgot to turn off your cell phone, and it rings, immediately turn it off. Do not answer the call.

在你走进餐厅之前，请关闭所有的电子设备，或者调成静音。如果你忘记关手机，来电话时你要马上关掉手机，不要去接电话。

2. Seating.

Avoid sitting at the head of a table (where the seats are fewest) unless you are a distinguished guest. Don't be the first to sit down, and when adjusting your chair, make sure to pick the chair up and move it in gently, rather than pulling it along the ground.

除非你是对方请来的贵宾，否则不要坐在桌子的最里头(里面的座位数量比较少)。不要第一个坐下，如果你要挪动椅子的话，一定要把椅子搬起来并轻轻地挪动它，不要把它放在地面上拖着走。

3. Posture.

When seated, make sure that you are no farther than eight inches away from the table. Your legs should not be extended, so as to respect the personal space of others. Forearms can either be placed on the table or in your lap. Always sit up straight.

你坐下以后,要保持你的身体在离桌子 8 英寸之内的距离。你的腿不要伸出去,这是极其不尊重他人的行为。前臂可以放在桌子上也可以放在腿上,要坐直。

4. Before you dig in...

Make sure that all others at the table are seated and served before you begin eating.

在你开始吃东西之前,要确保其他人都已经就座而且也开始吃了。

5. Silverware.

As a rule of thumb, silverware/utensils should be used in an outside-in order. When finished with a dish, find a place on that dish to neatly set your used utensils. Do not recycle these utensils if there are more arranged for you.

一般来说,银质的餐具(刀叉等),应该按照从外往里的顺序使用。吃完一道菜之后,在那道菜的盘子(或碗)里找个地方整齐地摆放你用过的餐具。如果餐具足够用,请不要重复使用一套餐具。

6. Eat, then talk.

A big no-no at formal dinner is talking while eating. Make sure that you finish the food in your mouth before communicating with others.

在正式餐会上的一大禁忌就是边吃边说,你在和别人交谈之前一定要先把嘴里的食物咽下去。

7. Being courteous.

Always thank the wait staff and cook (if present at the table) for a job well done. If you need to leave the table, excuse yourself from the group or at least with the people sitting next to you.

当服务员和厨师出现在你们的餐桌旁时,要感谢他们的优质服务。如果你需要离开餐桌,要跟这一桌的人解释一下,至少要跟坐在你旁边的人说一声。

8. Take your time.

Unless otherwise noted, dinner is not a competition to see who can finish the most food in the shortest amount of time. Pace yourself, and enjoy the company at the table.

如果没有人特别说明,餐会不是要比赛谁用最短的时间吃掉最多的食物。所以,你要控制你的进食速度,并且好好享受这个进餐的过程。

9. Be civilized.

Burping, farting, clipping your nails, picking your nose, and spitting are all unacceptable behavior at the dinner table.

打嗝、放屁、剪指甲、挖鼻子和吐痰等所有不文明的举止都绝对不可以在餐桌上发生。

# Example

## ✉ Dialogue 1

A: For our lunch meeting with the investors, do we have to make a reservation at the restaurant or do we just show up?
B: Usually for lunch, we don't have to reserve a table, they should allow walk-ins. But to be on the safe side, I'll order a table for half-past twelve. Will that suit your schedule?
A: I've arranged to meet them at the restaurant at twelve. Can you make the reservation a little earlier? If we start earlier, it will give us more time for a longer lunch.
B: Are you planning on treating the investors to a full-course meal?
A: Yes, we'll start with appetizers, follow with a soup and salad course, then main dishes of prime rib or cordon bleu chicken, and finish up with a delicious rich dessert of some sort.
B: That'll be pretty heavy for a mid-day meal, don't you think?
A: As long as we stay away from anything alcoholic, we should be okay.
B: With your prime rib and chicken choices, you'd better hope nobody's vegetarian.
A: We can make some special arrangement if we need to. After all, it's the company who is footing the bill.

## ✉ Dialogue 2

Waiter: Would you like to sit in smoking section, non-smoking section or whatever comes open first?
Tom: We prefer non-smoking section.
Waiter: Awfully sorry, there are no vacancies left now. Would you like to wait for a moment?
Tom: OK, we will wait a while.
(Ten minutes later)
Waiter: I'm sorry for making you wait for so long. Now there is a table available in non-smoking section. Please follow me.
Tom: Thank you.
Waiter: This is menu. Are you ready to order now?

· 32 ·

| | |
|---|---|
| Tom: | Sorry, we haven't decided yet. Could you please give us a little longer? |
| Waiter: | No problem. |
| Tom: | Well, I think I would like to have a shrimp cocktail and the tomato sausage soup first. How about you, Jones? |
| Jones: | The same for me please. |
| Waiter: | Yes, sir. |
| Jones: | What main dish would you like, Tom? |
| Tom: | Well, I'd like the sole, please. |
| Jones: | I'll have the sirloin steak. |
| Waiter: | Yes. What would you like to drink? |
| Jones: | Red wine for me, and you, Tom? |
| Tom: | I think I'll have beer. |
| Jones: | One goblet of red wine and one bottle of beer, please. |
| Waiter: | Would you like a dessert? |
| Jones: | What special kind of desserts do you have? |
| Waiter: | Lemon pie, hot cake in syrup, chocolate sundae and custard pudding. |
| Tom: | Well, I think we'll order after we finish the main course. |
| Waiter: | All right, I'll bring the soup right away. |

## Practice

Imagine that you are treating your partner who is new to Chinese cuisine at a Chinese restaurant. Practice the following conversation with the given points:

- ◆ Where would you like to sit?
- ◆ Make an introduction to some famous Chinese dishes.
- ◆ Ask for recommendations from a waiter.
- ◆ What would you like to order?
- ◆ How would you like to pay the bill?

## Tips

### ⊠ 英语就餐常用语

1. Do you like to go out eating?  想不想出去吃呢?
2. I'll treat you to dinner.  我想请您吃晚餐。

3. Let's go Dutch. 咱们各付各的吧。/ 咱们 AA 制吧。

4. Can I take a rain check? 您能改天再请我吗？

5. Is this seat taken? 这位子有人坐吗？

6. Have you made a reservation? 您预定了吗？

7. I am afraid the tables are all engaged at the moment, Sir/Madam, would you mind to wait for a while? 实在对不起，现在餐厅已经客满，请问您介意等一会儿吗？

8. Would you mind sharing a table? 您介意和别人同桌吗？

9. Smoking or non-smoking? 吸烟区还是非吸烟区？

10. May I take your order now? 请问我现在可以帮您点单吗？

11. My friend will be along shortly. 我的朋友一会就到。

12. I haven't decided yet. What do you recommend? 我还没想好。您推荐什么菜？

13. I am afraid this dish will take some more time to prepare. 对不起，这个菜需要一定的时间。

14. Are you used to the food here? 您习惯吃这儿的饭菜吗？

15. For here or to go? 在这儿吃还是带走？

16. My mouth is watering. 我在流口水了。

17. Enjoy your meal. 请慢慢享用吧。

18. I'm on a diet. 我正在节食。

19. No, thanks. I'm ready to burst now. 不了，谢谢，我的肚子快要胀破了。

20. This is a toast for … 这杯酒是敬……的。

21. How late are you open? 你们营业到几点？

22. May I settle your bill now? 请问现在可以为您结账吗？

23. Would you like to have one check or separate checks? 合开账单，还是单独付账？

24. May I pay by credit card? 我可以用信用卡付款吗？

25. Keep the change. 不用找了。

26. May I have the receipt? 可以给我开发票吗？

27. Mind your step and thank you for coming. 慢走，感谢您的光临。

28. How would you like your steak done, Sir/Madam, medium, medium well or well done? 请问您点的牛排喜欢几成熟，半熟、七成熟还是全熟？

29. What kind of dressing would you prefer with your steak? We have black paper sauce, red wine sauce, mushroom sauce, onion sauce. 您的牛排喜欢配什么汁呢？我们有黑椒汁、红酒汁、蘑菇汁和洋葱汁。

30. The buffet dinner is RMB 98 per person plus 15% service charge, and the service hour is from 18:00 till 21:00. 自助晚餐从 18:00 到 21:00，98 元人民币每位，加 15%的服务费。

 WRITING

# How to Write a Perfect Professional E-mail in English

## 如何写出一封完美的专业英文电子邮件

虽然电子邮件较商业书信而言有种不正式的感觉，但在商务电子邮件中仍应使用较正式的用语。电子邮件较快速并更有效率，但是你的客户或商业伙伴肯定不会喜欢太草率的邮件。记住以下一些简单的秘诀，将有助于提升你的英文电子邮件的专业水平。

1. 从问候开始

用问候语作为邮件开始是非常重要的，例如"Dear Lillian,"。根据你与收件人的关系亲近与否，你可能选择使用他们的姓氏来称呼他们而不是直呼其名，例如"Dear Mrs. Price,"。如果关系比较亲密，你就可以说"Hi Kelly,"。如果你和公司联系，而不是个人，你就可以写"To Whom It May Concern,"。

2. 感谢收件人

在回复客户的询问时，应该以感谢语开头。例如，如果有客户想了解你的公司，你就可以说"Thank you for contacting ABC Company."。如果此人已经回复过你的一封邮件，那就一定要说"Thank you for your prompt reply."或"Thanks for getting back to me."。要利用可能的任何机会感谢收信人，这样可令对方感到比较舒服，而且显得很有礼貌。

3. 表明你的目的

如果是你主动写电子邮件给别人，那就不要在开头写感谢的字句了，而应以你写此邮件的目的开头。例如，"I am writing to enquire about…"或是"I am writing in reference to…"。在电子邮件开头澄清你的来意非常重要，这样才能更好地引出邮件的主要内容。要注意语法、拼写和标点符号，同时避免冗长的句子，尽量让句子简短清楚。

4. 结束语

在结束邮件之前，应再次感谢收信人并加上一些礼貌语结尾。你可以写"Thank you for your patience and cooperation."或"Thank you for your consideration."并接着写"If you have any questions or concerns, don't hesitate to let me know."以及"I look forward to hearing from you."。

5. 结束

最后写上合适的结尾并附上你的名字。例如，"Best regards,"、"Yours faithfully,"、"Sincerely yours,"及"Thank you,"都是比较规范的用语。最好不要用"Best wishes,"或"Cheers,"之类的词，因为这些词都常用在非正式的私人邮件中。最后，在你发送邮件之前，最好再读一遍你的内容并检查其中有没有任何拼写错误，这样才能保证你发出的是一封真正完美的邮件。

# Example

Dear Mr. /Ms,

Mr. John Green, our General Manager, will be in Paris from June 2 to 7 and would like to come and see you, say, on June 3 at 2:00 p.m. about the opening of a sample room there.

Please let us know if the time is convenient for you. If not, what time you would suggest.

Yours faithfully,

David Hartley

---

✉ 尊敬的先生/小姐:

我们的总经理约翰·格林先生将于 6 月 2 日至 7 日在巴黎停留,希望能够于 6 月 3 日下午 2:00 拜访您并讨论有关在巴黎开样品房的事宜。

请告知这个时间对您是否方便。如不方便,另请建议具体时间。

您诚挚的

大卫·哈特利

---

Dear Mr. / Ms,

Thank you for your letter informing us of Mr. Green's visit during June 2-7. Unfortunately, Mr. Edwards, our manager, is now in London and will not be back until the second half of June. He would, however, be pleased to see Mr. Green any time after his return.

We look forward to hearing from you.

Sincerely yours,

Wang Lily

---

✉ 尊敬的先生/小姐:

谢谢来函告知我方 6 月 2~7 日格林先生的来访。不巧,我们的总经理艾德华先生现正在伦敦，他要到6月中旬才能回来。他回来后愿意在任何时间会见格林先生。

希望收到您的来信。

您诚挚的

王丽莉

## Practice

Dear Mr./Ms,
_____
_____
_____

Yours faithfully,
_____

☒ 尊敬的先生/小姐：

杰克巴伦先生，我们的人事主任对您向我们公司申请会计职位表示感谢，并请您于7月5日星期五的下午两点半来见他。

能否前来，请告知，多谢！

您诚挚的

Dear Mr./Ms,
_____
_____
_____

Sincerely yours,
_____

✉ 尊敬的先生/小姐：

谢谢昨日来信通知我面试，我将于要求的 7 月 5 日周五下午两点半到达，并带去我的证书及其他书面材料。

您诚挚的

# Tips

## ✉ 英文邮件常用语

1. 附件是……，请参阅。
   ◇ Please kindly find the attached…
   ◇ Attached please kindly find the…
   ◇ Enclosed is the …, please kindly find it.
   ◇ Attached you will find…

2. 我已经收到您的邮件。
   ◇ I have received your email (yesterday/last week/this morning…).
   ◇ Your email has been received.

3. 非常感谢您的回信。
   ◇ Thanks very much for your early reply.
   ◇ Thank you for your email.

4. 我不太明白您的意思。
   ◇ I do not quite understand what you meant.
   ◇ I have some problem understanding what you meant.

5. 您能具体解释一下么？您能再说得清楚一点么？
   ◇ Could you please put it in a clearer way?
   ◇ Could you please explain it in detail?
   ◇ Could you please further explain it?
   ◇ What do you mean exactly by saying…?

6. 如有问题，请随时和我联系。
   ◇ Contact me if you have any problem.
   ◇ If there is any uncertainty, feel free to contact me.

- Call me if you have any problem.

7. 一旦确认，我会立刻/第一时间通知您的。
- Once confirmed, I will let you know immediately.
- Once confirmed, I will notice you ASAP. （ASAP=as soon as possible，尽快）

8. 希望尽快得到您的回复。
- I'm looking forward to your early reply.
- Your promote reply will be greatly appreciated.
- Your early reply will be highly appreciated.

9. 感谢。
- Thanks a lot.
- Thank you so much for the cooperation!
- Thanks for your attention!
- Many thanks. （用于关系熟悉的同事或朋友间。）

10. 祝福。
- I hope everything goes well with you.
- I wish you all the best.
- Hope you have a good trip back.

## ✉ 英文邮件常见缩写

1. AOB    any other business    其他事

2. ASAP    as soon as possible    尽快

3. BCC    blind carbon copy    密送

4. BTW    by the way    顺便提一下

5. CC    carbon copy    抄送

6. Conf.    confidential    机密

7. FYI    for your information…    供您参考

8. IOW    in other words    换句话说

9. PS    postscript    附言

10. WRT    with regard to    关于

# UNIT3

## TOPICS

- What is computer software?
- The kinds of software.
- What is the application software?
- What is packaged software?
- What is custom-made software?
- What are basic tools?
- What is system software?
- Knowing some commonly used microcomputer operating systems.
- What is programming language?
- How to brainstorming?
- How to write application and recommendation letters?

 TEXT

# What Is Computer Software?

Software refers to the programs that are loaded onto a computer, or another name for programs. Programs are the **instructions** that tell a computer how to process data into the form you want. **In most cases**, the words software and programs are **interchangeable**.

There are two major kinds of software—application software and system software. You can think of application software as the kind you use and of system software as the kind the computer uses.

**Application Software**

Application software might be described as "**end-user**" software. Application software performs useful work on **general-purpose** tasks such as word and **spreadsheet** processing. Application software may be packaged or custom-made.

**Packaged** software refers to programs **prewritten** by professional programmers that are typically offered for sale on a **diskette**. There are over 12,000 different types of application packages available for **microcomputers** alone.

**Custom-made** software, or custom programs, is what all software used to be. Twenty years ago organizations hired computer programmers to create all their software. The programmer custom-wrote programs to instruct the company computer to perform whatever tasks the organization wanted. A program might compute payroll checks, **keep track of** goods in the warehouse, calculate sales commissions, or performs similar business tasks.

There are certain general-purpose programs that we call "**basic tools**". These programs are widely used in nearby all career areas. They are the kind of programs you have to

instruction [in'strʌkʃən] n. 指令, 教学, 说明

in most cases 在多数情况下, 往往

interchangeable [ˌintə'tʃeindʒəbl] adj. 可互换的

end-user [计] 终端用户

general-purpose ['dʒenərəl'pə:pəs] adj. 通用的

spreadsheet ['spred,ʃi:t] n. 电子制表软件, 电子表格

packaged ['pækidʒd] adj. 包装的, 成套的, 打包的

prewritten ['pri'ritən] adj. 预先写的

diskette ['disket] n. [计] 软盘

microcomputer['mækrəukəm'pju:tə] n. 微型计算机

custom-made ['kʌstəm,med] adj. 定做的, 订制的

keep track of 跟上……的进展, 记录

basic tool 基本工具

know to be considered computer competent. The most popular so-called basic tools include:

- Word processing programs, used to prepare written documents.
- Electronic spreadsheets, used to analyze and **summarize** data.

summarize ['sʌməraiz] v. 概述,摘要而言

- Database managers, used to organize and manage data and information.
- Graphics programs, used to visually analyze and present data and information.
- Communication programs, used to transmit and receive data and information.
- **Integrated programs**, which combine some or all of these applications in one program.

integrated program 集成程序

## System Software

The user **interacts** with application software. System software enables the application software to interact with the computer. System software is "**background**" software, including programs that help the computer manage its own internal resources.

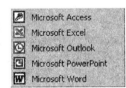

interact [ˌintə'rækt] vi. 相互作用,相互影响,互动

background ['bækgraund] n. 背景,经历,后台

The most important system software program is the operating system, which interacts between the application software and computer. The operating system handles such details as running ("**executing**") programs, storing data and programs, and processing ("**manipulating**") data. System software frees users to **concentrate on** solving problems rather than on the **complexities** of operating the computer.

execute ['eksikju:t] vt. 执行,实行
manipulate [mə'nipjuleit] vt. 操纵,操作,控制
concentrate on 集中于,专心于
complexity [kəm'pleksiti] n. 复杂,复杂性,复杂的事物

Microcomputer operating systems change as the machines themselves become more powerful and outgrow the older operating systems. Today's computer competency, then, requires that you have some knowledge of the following most popular microcomputer operating systems:

- DOS, the standard operating system for personal computers manufactured by International Business Machines(IBM) and IBM-**compatible** microcomputers.

- Windows, which initially, is not an operating system, but an environment that extends the **capability** of DOS.

- Windows NT, a powerful operating system designed for powerful microcomputers.

- OS/2, the operating system developed for IBM's more powerful microcomputers.

- Macintosh operating system, the standard operating system for Apple Corporation's Macintosh computers.

- UNIX, an operating system **originally** developed for minicomputers. UNIX is now important because it can run on many of the more powerful microcomputers.

**Programming Language**

A programming language is an **artificial** language designed to express computations that can be performed by a machine, particularly a computer. Programming languages can be used to create programs that control the behavior of a machine, to express **algorithms** precisely, or as a mode of human communication.

The earliest programming languages **predate** the invention of the computer, and were used to direct the behavior of machines such as playing pianos. Thousands of different programming languages have been created, mainly in the computer field, with many more being created every year. Most programming languages describe computation in an **imperative** style, i.e., as a sequence of commands, although some languages, such as those that support

compatible [kəm'pætəbl] adj. 一致的, 兼容的  n. 兼容

capability [ˌkeipə'biliti] n. 能力, 性能, 容量

originally [ə'ridʒənəli] adv. 原本, 起初

artificial [ˌɑːti'fiʃəl] adj. 人造的, 人工的

algorithm ['ælgə.riðəm] n. 算法

predate [priː'deit] v. 在日期上早于

imperative [im'perətiv] adj. 紧要的, 必要的, 祈使的

functional programming or logic programming, use alternative forms of description.

A programming language is usually split into the two components of **syntax** (form) and **semantics** (meaning) and many programming languages have some kind of written specification of their syntax and/or semantics. Some languages are defined by a **specification** document, for example, the C programming language and BASIC programming language are specified by an ISO Standard.

syntax ['sintæks] n. 句法
semantics [si'mæntiks; sə'mæntiks] n. 语义学, 符号学, 语法
specification [.spesifi'keiʃən] n. 规格, 详述

# EXERCISES

Ⅰ. Match the terms and the interpretations.

1. Operating System  (a) Is computer software designed to help the user to perform specific tasks. Examples include enterprise software, accounting software, office suites, graphics software and media players.

2. Application Software  (b) Is a program or collection of programs that enable a person to manipulate visual images on a computer.

3. Graphics Software  (c) Is a set of programs that manage computer hardware resources and provide common services for application software.

4. System Software  (d) Is a step-by-step procedure for calculations. It is used for calculation, data processing, and automated reasoning.

5. Algorithm  (e) Designed to operate the computer hardware and to provide a platform for running application software.

Ⅱ. Are the following statements True (T) or False (F)?

1. (    ) Sometimes we can use the words software instead of program.

2. (    ) Microsoft Word is custom-made software.

3. (    ) The most important system software is Internet Explorer.

4. (    ) Linux is one of the Operating System.

5. (    ) C Language can't be used to develop system software.

6. (    ) In the beginning, there were a large variety of applications built on Linux.

7. (    ) The Calculator which installed along with Windows is system software.

8. (    ) Programming is the act of creating a computer program, a concrete set of instructions for a computer to carry out.

9. (    ) Machine language program is running faster than Java program (same program).

10. (    ) Object-oriented languages are based on an approach to programming that uses objects.

III. Translate the following words and phrases into Chinese.

1. Spreadsheet processing          _____

2. Packaged software               _____

3. Custom-made software            _____

4. Database manager                _____

5. Graphics programs               _____

6. Algorithm                       _____

7. Processing data                 _____

8. IBM-compatible microcomputers   _____

9. Extends the capability of DOS   _____

10. Specification document         _____

IV. Translate the following Chinese statements into English.

1. 操作系统告诉计算机怎样完成最基本的功能和如何解释用户指令。

2. 不同的操作系统是针对不同的计算机设计的，比如 Windows 运行于个人电脑上，而 MVS 运行于大型机上。

3. Windows 用户不必记住和输入那些在 DOS 上使用的复杂的命令。

___

4. 许多人认为 Windows XP 远远不如所期望的那么好。

___

5. 你如何决定哪种程序设计语言更适合你。

___

Ⅴ. Fill in each of the blanks with one of the following words.

**1. System Software**

*supervises    standard routines    convert    maintenance*
*loader programs    among    categories*

Software systems can be subdivided into six different _____ as follows:
- Language processors that _____ programs from user-oriented languages to machine language.
- Library programs that provide _____ for the application programmer.
- Utility programs that facilitate the communication _____ computer components and between computers and users.
- _____ that make it easier to read various programs into memory.
- Diagnostic programs that facilitate the _____ of the computer.
- The operating system that _____ all other programs and controls their execution.

**2. The Operating System**

*platform    operation    available    efficiency    cost    application*

The Operating system is a set of system programs that, when executed, controls the _____ of the computer. Sophisticated operating systems increase the _____ and consequently decrease the _____ of using a computer. Operating systems provide a software _____, on top of which other programs, called _____ programs, can run. The application programs must be written to run on top of a particular operating system. Your choice of operating system, therefore, determines, to a great extent, what applications you can run. For PCs, the most popular operating systems are DOS, OS/2, and Windows, but others are _____, such as Linux.

**3. Interacting with the OS**

*commands    interact    executed    clicking    processor*

As a user, you normally _____ with the operating system through a set of _____. For example, the MS-DOS operating system contains commands such as COPY and RENAME for copying files and changing the names of files, respectively. The commands are accepted and _____ by a par tof the operating system called the command _____ or command line interpreter. Graphical user interfaces allow you to enter commands by pointing and _____ at objects that appear on the screen.

## SUPPLEMENTARY

### 9 Common Windows 7 Problems

We like Windows 7: it's faster than Vista, makes better use of your system resources, is packed with interesting features, and looks great. But that doesn't mean it's perfect, of course. If you've moved to Windows 7 recently then you might have noticed various upgrade problems, interface issues and features that seem to have disappeared entirely, among many other complications with the new system.

Don't despair, though—while these problems can be really frustrating, answers are beginning to appear. We've uncovered some of the best and most effective solutions around (For space reasons, only part of the solutions are listed here), so follow our guide and your Windows 7 installation will soon be back on track.

1. Vista upgrade hangs at 62%

Windows 7 can start causing problems before it's even installed, as many people report their upgrade hangs forever at 62%. Which is annoying?

2. DVD drive not found

In some cases your DVD drive may not be found by Windows 7, even if it's visible in the BIOS and using the standard driver.

3. Windows 7 themes change your custom icons

Windows 7 has some spectacular new themes—there's a great selection at the Microsoft site—but installing them can have one annoying side-effect. If you've previously changed a system icon like Computer or the Recycle Bin then that could disappear, replaced by the equivalent icon from the theme pack.

To prevent this, right-click an empty part of the desktop, select Personalize→Change Desktop Icons, clear the "Allow themes to change desktop icons" box and click OK. Your icons will now be preserved, and the only way to change them will be manually, from the same Desktop Icons dialogue.

## 4. Taskbar problems

We like the new Windows 7 taskbar, but many people seem less than impressed with the new approach to taskbar buttons, finding it difficult to tell at a glance whether an icon is a running application or a pinned shortcut. If this sounds like you then there's an easy way to restore more standard taskbar buttons, though—right-click the taskbar, select Properties, and set Taskbar Buttons to "Never combine" or "Combine when taskbar is full".

## 5. Missing Explorer folders

Click Start→Computer in Windows 7 and you'll find system folders like Control Panel and the Recycle Bin are no longer displayed in the left-hand Explore pane. This seems like a backward step to us, but there's a quick solution. Click Tools→Folder Options, check "Show all folders", click OK and all your top-level system folders will reappear.

## 6. Missing applets

Windows 7 installs quickly and takes up less hard drive space than you might expect, but in part that's down to cheating—Mail, Movie Maker, Photo Gallery and other applets are no longer bundled with a standard Windows installation. Instead you must download the programs you need from the Windows Live Essentials site.

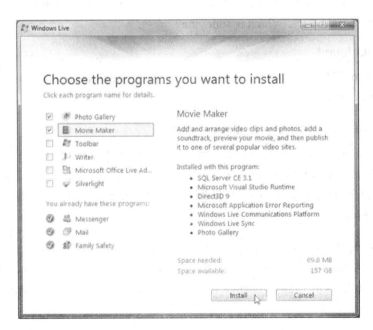

## 7. Too many mini dumps

By default Windows 7 now keeps the last 50 minidump files (memory images saved when your PC crashes). If you're keen on using dump files to troubleshoot crashes then this is good news, but if you've no interest in that kind of advanced debugging then minidumps is just a waste of your valuable hard drive space. In which case you should run REGEDIT, browse to HKEY_LOCAL_MACHINE\SYSTEM\CurrentControlSet\Control\CrashControl, and set MiniDumpsCount to 1. Windows will only now keep the last dump file and you'll free up a little hard drive space.

## 8. HP Multifunction Printer problems

If you've an HP multifunction printer with its "Full Feature Software solution" or "Basic Driver solution" installed then, after upgrading to Windows 7, you may find the printer stops working. Press the buttons on the front of the printer and nothing will happen; launch the software manually and you'll see reports that it can't connect to your hardware.

The problem is that a few files and Registry entries have been lost in the migration to Windows Vista, and even reinstalling the original HP software won't help. Fortunately there's a new version of HP Solution Center that should get everything working again, though, and you can find out more about it at the HP support site.

## 9. Hidden extensions

And, of course, no list of Windows annoyances would be complete without a mention of Explorer's default settings, which even in Windows 7 remain to hide file extensions, as well as system files and folders.

To fix this, launch Explorer and click Tools→Folder Options→View.

Clear the "Hide extensions for known file types" to show file extensions, reducing the likelihood that you'll accidentally double-click on virus.txt.exe in future.

And as long as there are no novice users on your system who might go poking around in Explorer, we'd also choose to "Show hidden files and folders" as well as clear the "Hide protected operating system files" box. It's often important to see these files when you're troubleshooting, or following problem-solving instructions from someone else.

# CONVERSATION

## Brainstorming

## 头 脑 风 暴

头脑风暴法又称智力激励法,是现代创造学奠基人——美国的斯本提出的,是一种创造能力的集体训练法。头脑风暴法力图通过一定的讨论程序与规则来保证创造性讨论的有效性,由此,讨论程序构成了头脑风暴法能否有效实施的关键因素。从程序上来说,组织头脑风暴的关键在于以下几个环节。

### 1. 确定议题

一个好的头脑风暴法是从对问题的准确阐明开始的。因此,必须在会前确定一个目标,使与会者明确通过这次会议要解决什么问题,同时不要限制可能的解决方案的范围。一般而言,比较具体的议题能使与会者较快产生设想,主持人也较容易掌握;比较抽象和宏观的议题引发设想的时间较长,但设想的创造性也可能较强。

### 2. 会前准备

为了使头脑风暴畅谈会的效率较高,效果较好,可在会前做一点准备工作。如收集一些资料预先给大家参考,以便与会者了解与议题有关的背景材料和外界动态。就参与者而言,在开会之前,对于要解决的问题一定要有所了解。会场可做适当布置,座位排成圆环形的环境往往比教室式的环境更为有利。此外,在头脑风暴会正式开始前还可以出一些创造力测验题供大家思考,以便活跃气氛,促进思维。

### 3. 确定人选

与会者一般以 8~12 人为宜,也可略有增减(5~15 人)。与会者人数太少不利于交流信息,激发思维,而人数太多则不容易掌握,并且每个人发言的机会相对减少,也会影响会场气氛。只有在特殊情况下,与会者的人数可不受上述限制。

### 4. 明确分工

要推选一名主持人及 1~2 名记录员(秘书)。主持人的作用是在头脑风暴畅谈会开始时重申讨论的议题和纪律,在会议进程中启发引导,掌握进程。如通报会议进展情况,归纳某些发言的核心内容，提出自己的设想，活跃会场气氛，或者让大家静下来认真思索片刻再组织下一个发言高潮等。记录员应将与会者的所有设想都及时编号，简要记录，最好写在黑板等醒目处，让与会者能够看清。记录员也应随时提出自己的设想，切忌持旁观态度。

### 5. 规定纪律

根据头脑风暴法的原则，可规定几条纪律，要求与会者遵守。如要集中注意力积极投入，不消极旁观；不要私下议论，以免影响他人的思考；发言要针对目标，开门见山，不要客套，也不必做过多的解释；与会之间相互尊重，平等相待，切忌相互褒贬等等。

### 6. 掌握时间

会议时间由主持人掌握，不宜在会前定死。一般来说，以几十分钟为宜。时间太短与会者难以畅所欲言，太长则容易产生疲劳感，影响会议效果。经验表明，创造性较强的设想一般要在会议开始10分钟~15分钟后逐渐产生。美国创造学家帕内斯指出，会议时间最好安排在30分钟~45分钟之间。倘若需要更长时间，就应把议题分解成几个小问题分别进行专题讨论。

# Example

## Dialogue 1

Jane: Tom! Your presentation was awesome! I'm really impressed!
Tom: thank you, Jane! Our group members had a brainstorm last week! So the presentation was so successful.
Jane: Brainstorm? What's that?
Tom: That means, our group members put our heads together and to discuss.
Jane: Oh! I think I know what you are talking about! You guys had a brainstorming session.
Tom: That's right. Brain, b-r-a-i-n, storm, s-t-o-r-m, put them together that is the word brainstorm
Jane: Exactly! Brainstorm basically means to put your heads together in order to come up with good ideas.
Tom: Hey Jane! Susan's birthday is coming soon. Let's put our heads together and figures out what we can give her for her birthday.
Jane: Good idea!

## ⮞ Dialogue 2

A: We've got to come up with a way to solve this problem! No one is getting all their work done during the work day. We need to do some brainstorming to come up with a time-management solution for our office.

B: Well, what do you suggest? It's not that we are all wasting a lot of time. Fact is, we're understaffed. People just have too much to do.

A: So you think we should hire some new people? That is one possible solution. What else could be the problem?

B: Maybe we do have some problems with time-management. If people were always on time to work, they might get more done in the morning before lunch.

A: That's true. Also, maybe we could shorten the lunch break from one hour to forty five minutes. That would add a little time to everyone's day.

B: Fifteen minutes would be good, but I wonder if it would make a big impact on the employees' output. I still think it's a problem with too much workload.

A: We ended up right where we started. Maybe it is time to look for some temp workers to help with a few projects. That would lighten everyone's load.

## Practice

Imagine that you and your colleagues are in a meeting, brainstorming "Which language is the best choice in developing the new HR project?"

- ✧ Project Manager makes a brief introduction to the project and the brainstorming topic.

- ✧ Tom: C is the basis of any other language. We have only 6 people in our project team, but almost all of our programmers can use C.

- ✧ Mary: Java is open-source, and very popular now.

- ✧ Jim: C# is better, because it's Microsoft's invention, and the Integrated Development Environment makes everything easier.

- ✧ Discuss the advantages and disadvantages of each programming language.

- ✧ Reach an agreement.

# Tips

## ⊠ 头脑风暴常用语

1. That's true. We can sit down and discuss, brainstorming with our team members. All of us can get some inspiration from different people's ideas. 对啊，我们坐下来讨论，和队员来一个头脑风暴。我们每个人都能从不同人的不同想法中得到一些灵感。

2. Yes, we can be more efficient if we make the best use of everybody's idea. 是的，如果能充分利用每个人的想法，我们就能更有效地做事。

3. We can get a better solution with discussion in shorter time. 我们通过讨论在较短的时间得到更好的解决方法。

4. Sounds quite reasonable. But everybody has his own ideas. 听起来很有道理，但是每个人都有自己的想法。

5. Oh, you are always so professional. Do you have any good ideas? 噢，你总是那么专业。有什么好的意见吗？

6. Thanks for saying that. I think if we have a better teamwork, it will be much better. 谢谢！我想如果我们有一个很好的团队合作，就会好很多。

7. That's a good idea. Let's give it a try. 是个好主意。我们试一试吧。

8. In that case, all of us can make the progress at the same time. 如果那样的话，我们就能共同进步了。

9. I hope my advice will be of some use to you. 我希望我的建议能对你有帮助。

10. You are so great! I envy you. 你真了不起！我很佩服你。

11. Be confident with yourself. I am sure you'll do fine. 你也自信些，我相信你能做得很好的。

12. Oh, I am at the end of my wits and running out of steam. 噢，我江郎才尽了。

13. The creative works need inspiration. 有创造性的工作是需要灵感的。

# WRITING

# How to Write Application and Recommendation Letters

# 如何写求职信及推荐信

### 1. 求职信(Application Letter)

求职信是申请具体职位的信件，一封好的求职信对于取得某职位可能会取得事半功倍的效果。一封好的求职信应该包括下面一些要点：

- 说明你要申请的职位和你是如何得知该职位的招聘信息的。
- 说明你如何满足公司的要求，陈述个人技能和个性特征。
- 表明你希望迅速得到回音，并标明与你联系的最佳方式。
- 感谢对方阅读并考虑你的应聘。每封求职信应以针对雇主而精心设计，以此表明你明白该公司的需要。
- 求职信还应包括你所取得的成果及解决的问题的事例，这些事例与你所申请的工作类型相关。
- 求职信应是寄给有职位的某一特定的人，使用高档纸书写，仔细校对，避免打字或语法方面的错误，要自存副本档案。

### 2. 推荐信(Recommendation Letter)

推荐信是向有关单位或个人推荐有关人才的专用书信。推荐信可以是个人写给个人、个人写给单位或单位写给单位的。推荐信一般是由第三者写给对方，也有向某单位、部门自荐的。推荐信一般包括称谓、正文、结尾、落款和日期几部分。其中正文应该包括被推荐者的基本情况，包括姓名、性别、年龄、业务水平、工作能力、身体状况等。另外，正文中还应该包括推荐的理由，理由最好写得具体、充分，并还要写明推荐者和被推荐者的关系。

推荐信要客观全面地评价被推荐人（或自己），要求如实地介绍被推荐人（或自己）的优点，客观地叙述推荐的理由，以便供有关人员参考。

# Example

**Application Letter**

Dear Sir,

I am writing to apply for the position as an English teacher that you recently advertised in Sichuan Daily. I take keen interest in the post because I find that my major and experiences well meet the requirements you stated in the advertisement.

Being interested in English teaching, I pursued my graduate study in the direction of teaching methodology in Sichuan International Studies University, and got a Master's Degree in 2003. I was a top student through the three academic years, as can be shown in the enclosed resume and reports. After graduation, I ever taught English in a Xi'an high school. As Sichuan is my hometown I love very much, I have decided to move back and so I venture to apply for the position in your school.

If I were favored with an interview, I would be most grateful. Please contact me at 13573889787. Thank you for your consideration.

Best wishes.

Yours sincerely,

Li Ming

求 职 信

你在《四川日报》上看到了某学校要招聘英语老师的广告，因此写了这封求职信，申请到该校工作。你于2003年获得了四川外语学院的英语教育硕士学位，在校的三年中一直是尖子生。毕业后你在西安一所高中教学。由于你非常喜欢自己的家乡四川，所以现在想应聘四川的这所学校。

**Recommendation Letter**

Dear Mr. Wang,

I am very sorry to tell you that I am going to graduate this June and cannot go on with my job as a tutor of your daughter. It has really been a pleasant experience to teach your daughter English as she is such a lovely and smart girl. Here I take great pleasure in recommending to you my friend Lily who is a sophomore majoring in English in my university. She is particularly willing to take the part-time job of an English tutor when she knows about your daughter.

Lily is an excellent student. Especially her spoken English is both fluent and proficient, which can positively influence the person speaking with her. Moreover, as a lively, cheerful and easy-going girl, she is good at communicating with others.

Therefore, I am confident that she is highly competent for the job and will help your daughter make further progress in English.

Sincerely yours,

Li Ming

推 荐 信

由于你即将于今年六月份毕业，所以不能继续做王先生女儿的英语家庭教师了，于是你写这封推荐信推荐你的朋友莉莉，她是你所在大学的英语专业大二的学生。当听说王先生的女儿聪明伶俐、口语又好时，莉莉很希望能做这份兼职英语教师的工作。

# Practice

Dear Mr. Smith,

_____

_____

_____

_____

Sincerely yours,

## 求 职 信

你即将毕业于西安交通大学计算机软件专业，在网站上看到"West Creative"软件公司的招聘信息，于是写了一份求职信，希望能加入该公司，从事软件开发或软件测试工作。

> To whom it may concerns,
>
> _____
> _____
> _____
> _____
> _____
>
> Sincerely yours,

## 推 荐 信

你是常州信息学院经贸管理学院的学生辅导员，推荐你的学生张亮到蓝鑫外贸公司工作。张亮的专业是国际贸易，他的英语听说能力非常好，他在校是学生会副主席，人际交往和沟通能力也都很棒。

## Tips

### 推荐信和求职信常用语

1. Mature, dynamic and honest. 思想成熟、精明能干、为人诚实。

2. Excellent ability of systematical management. 有极强的系统管理能力。

3. Ability to work independently, mature and resourceful. 能够独立工作，思想成熟、应变能力强。

4. A person with ability plus flexibility should apply. 需要有能力及适应力强的人。

5. A stable personality and high sense of responsibility are desirable. 个性稳重、具高度责任感。

6. Work well with a multi-cultural and diverse work force. 能够在多元化的环境下出色地工作。

7. Bright，aggressive applicants. 反应快、有进取心的应聘者。

8. Ambitious attitude essential. 有雄心壮志。

9. Initiative，independent and good communication skill. 积极主动、独立工作能力强，并有良好的交际技能。

10. Willing to work under pressure with leadership quality. 愿意在压力下工作，并具领导素质。

11. Willing to assume responsibilities. 应聘者须勇于挑重担。

12. Mature，self-motivated and strong interpersonal skills. 思想成熟、上进心强，并具极丰富的人际关系技巧。

13. Energetic，fashion-minded person. 精力旺盛、思想新潮。

14. With a pleasant mature attitude. 开朗成熟。

15. Strong determination to succeed. 有获得成功的坚定决心。

16. Strong leadership skills. 有极强的领导艺术。

17. Ability to work well with others. 能够同他人一道很好地工作。

18. Highly-motivated and reliable person with excellent health and pleasant personality. 上进心强又可靠者，并且身体健康、性格开朗。

19. The ability to initiate and operate independently. 有创业能力，并能独立地工作。

20. Strong leadership skill while possessing a great team spirit. 有很高的领导艺术和很强的集体精神。

21. Be highly organized and efficient. 工作很有条理，办事效率高。

22. Willing to learn and progress. 肯学习进取。

23. Good presentation skills. 有良好的表达能力。

24. Positive active mind essential. 有积极、灵活的头脑。

25. Ability to deal with personnel at all levels effectively. 善于同各种人员打交道。

26. Have positive work attitude and be willing and able to work diligently without supervision. 有积极的工作态度，愿意和能够在没有监督的情况下勤奋地工作。

27. Young，bright，energetic with strong career-ambition. 年轻、聪明、精力充沛，并有很强的事业心。

28. Good people management and communication skills. Team player. 有良好的人员管理和交际能力，能在集体中发挥带头作用。

29. Able to work under high pressure and time limitation. 能够在高压力下和时间限制下进行工作。

30. Be elegant and with nice personality. 举止优雅、个人性格好。

31. With good managerial skills and organizational capabilities. 有良好的管理艺术和组织能力。

32. The main qualities required are preparedness to work hard, ability to learn, ambition and good health. 主要必备素质是吃苦耐劳、学习能力强、事业心强和身体棒。

33. Having good and extensive social connections. 具有良好而广泛的社会关系。

34. Being active, creative and innovative is a plus. 思想活跃、首创和革新精神尤佳。

35. With good analytical capability. 有较强的分析能力。

36. In reply to your advertisement in today's (newspaper), I respectfully offer my services for the situation. 拜读今日 xx 报上贵公司的广告，本人特此备函应征该职位。

37. Shall you need an experienced desk clerk for your hotel next summer? 贵酒店明年暑期是否需要一名有经验的柜台部职员？

38. I have been for over five years in the employ of an exporting company. 本人曾经前后五年受雇于出口贸易公司。

39. My interest in the position of masonry supply manager has prompted me to forward my resume for your review an consideration. 由于我对砌体供应经理这个职位非常感兴趣，所以我向您递交了我的简历供您审阅和考虑。

40. Are you currently seeking a security specialist to maintain or upgrade the security of your organization website? If so, I would like to apply for the position. 您是否正在寻找一个安全专家维护或升级你们组织的网站？如果是这样，我想申请这个职位。

# UNIT 4

## TOPICS

- What's Software Engineering?
- What's the principles in Software Engineering?
- What's modifiability in Software Engineering?
- What's understandability in Software Engineering?
- What's adaptability in Software Engineering?
- What's the target for Software Engineering?
- What's reliability in software engineering?
- Telephone booking skills.
- How to write IOU and receipt?

# Software Engineering

Software Engineering (Software Engineering, referred to as SE) is a research discipline to build and maintain an effective, practical and high quality software engineering method. It relates to programming languages, databases, software development tools, platform, standards, and design patterns.

In modern society, the software used in many ways. The typical software such as E-mail, **embedded systems**, human-machine **interface**, office suites, operating systems, compilers, databases, games, etc. Meanwhile, various industries almost all computer software applications, such as industry, agriculture, banking, aviation, government departments. Promote economic and social development of these applications, making people work more efficiently, while improving the quality of life.

embed system[计] 嵌入式系统
interface ['intəfeis] n. 界面；[计]接口；交界面 vt. (使通过界面或接口)接合，连接；[计]使联系 vi. 相互作用(或影响)；交流，交谈

**Defined**

Recognized a definition: software engineering research and application of systematic, standardized, **quantitative** process to develop and maintain the software and how to prove the right management techniques and **time-tested** able to get the best technology and methods together.

quantitative ['kwəntitətiv] adj. 定量的；数量(上)的
time-tested ['taim'testid] adj. 经受时间考验的，久经试验的

**Target**

The goal of software engineering is: under the premise of a given cost, schedule, development can be **modified**, the validity, reliability, understandability, maintainability, reusability of software engineering and adaptability, portability, traceability and interoperability and to meet user needs software products. The pursuit of these goals will help to improve software product quality and development

target ['ta:git] n. (服务的)对象；目标；(射击的)靶子；目的 vt. 瞄准，把……作为攻击目标
modifiy ['mɔdifai] vt. 更改，修改

efficiency, reduce maintenance difficulties.

- **Modifiability**

Allow the system to modify without increasing the original system complexity. It supports software debugging and maintenance, is a difficult target.

- **Efficiency**

The software system can most effectively use the computer resources of time and space resources. A variety of computer software are all the system time/space overhead as an important technical indicators to measure software quality. Many occasions, in the pursuit of time effectiveness and space **efficiency**, conflict, then had to sacrifice time efficiency in exchange for space efficiency or sacrificing space efficiency for time validity.

- **Reliability**

To **prevent** the software system failure caused by imperfect concept, design and structure, has to restore the software system failure due to improper operation. For real-time embedded computer systems, reliability is a very important goal. If reliability is not assured, if there are problems could be catastrophic, **consequences** will be unimaginable. In software development, coding and testing process reliability in an important position.

- **Understandability**

The system has a clear structure, a direct reflection of the needs of the problem. The intelligibility help control the complexity of software systems and support software maintenance, **migration** or reuse.

- **Maintainability**

Software products delivered to users, it can be modified in order to correct potential errors, improve performance and other attributes, to make the software adapt to environmental changes, and so on. Because the software is a logical product, as long as the user needs, it can be indefinitely

---

modifiability[,mɔdi,faiə'biliti] n. 可修改性

efficiency [i'fiʃənsi] n. 效率, 效能; 实力, 能力; [物]性能; 功效 [复]efficiencies

prevent [pri'vent] vt. 预防; 阻碍; 阻止; [宗教]引领 vi. 阻挠, 阻止

consequence ['kɔnsikwəns] n. 结果, 成果; [逻]结论; 重要性; 推论

migration [mai'greʃiən] n. 迁移, 移居

using it, the software maintenance is inevitable. Software maintenance costs account for a large proportion of software development costs.

- **Reusability**

Concept or function relatively independent of one or a group of related modules defined as a soft component. The soft parts in a variety of occasions, the application level as the reusability of components. **Reusable** software components and some modification to be used directly, and some need to modify and then use. Reusable soft parts with a clear structure and explanatory notes should have the correct coding and a lower time/space overhead. In the broader sense of understanding, the reusability of software engineering should also include: the reuse of application projects, specifications (also known as the Statute), reuse, design reuse, the reuse of concepts and methods, and so on. In general, the higher the level reuse, the greater the benefits.

reusable[,riːˈjuːzəbl] adj. 可再度使用的；可多次使用的

- **Adaptability**

Different system constraints, the software allows the user needs to the degree of difficulty to be met. The adaptability of the software should be used widely popular programming language code to run in the widely popular operating system environment, the use of standard **terminology** and format for writing documentation.

terminology[,təːmiˈnɔlədʒi] n. 术语；术语学

- **Portability**

Ease of software from one computer system or the environment to move to another computer system or the environment. In order to obtain a relatively high portability, the software design processes usually support general-purpose programming language and runtime environment. Low-level (physical) characteristics of part dependent on computer systems, such as the build system of the target code generation, should be relatively independent, and concentrated. In this way, and processor-independent part can be ported to other systems.

- **Traceability**

Software requirements, software design, program forward track, according to the procedures, software design, software requirements, ability to reverse track. Software traceability relies on the integrity, consistency and comprehensibility of the documents and programs of the various stages of software development. Reduce the complexity of the system will improve the traceability of the software.

- **Interoperability**

Multiple software elements communicate with each other and collaborative tasks. In order to achieve interoperability, software developers usually have to follow certain standards, and support to facilitate the compromise standard of environment for the software elements between interoperable. Interoperability in a distributed computing environment is particularly important.

- **Process**

To produce a final can meet the needs and the steps required to achieve the goal of the project's software products. The software engineering process includes the development process, operations, and maintenance procedures. They cover the needs, design, implementation, verification and maintenance and other activities.

- **Principle**

Software engineering principle is to focus on engineering design, engineering support, and must follow the principles of project management in the software development process. The framework of the software engineering tells us that the goal of software engineering is the availability, accuracy, and cost-effective; the implementation of a software engineering to select the appropriate development paradigm, appropriate design methods, to provide high quality engineering support, to implement effective management of the development process; software engineering activities including requirements, design, implementation, to confirm

traceability[treisə'biləti] n. 可追踪性

principle['prinsəpl; 'prinsəpəl] n. 原理；原则；主义；信念

and support activities, each activity can be based on the specific software engineering, development paradigm, design methods, support the process and process management.

## EXERCISES

Ⅰ. Match the terms and the interpretations.

1. Software Engineering (SE)
2. Modifiability
3. Efficiency
4. Maintainability
5. Adaptability

(a) Different system constraints, the software allows the user needs to the degree of difficulty to be met.

(b) Software products delivered to users, it can be modified in order to correct potential errors, improve performance and other attributes, to make the software adapt to environmental changes, and so on.

(c) Allow the system to modify without increasing the original system complexity.

(d) The software system can most effectively use the computer resources of time and space resources.

(e) A research discipline to build and maintain an effective, practical and high quality software engineering method. It relates to programming languages, databases, software development tools, platform, standards, and design patterns.

Ⅱ. Are the following statements True (T) or False (F)?

1. (　) Software Engineering is not a study of engineering methods to build and maintain effective, practical and high-quality software disciplines.

2. (　) All the system time/space overhead as an important technical indicators to measure software quality

3. (　) The system has a clear structure, a direct reflection of the needs of the problem.

4. (　) To produce a final can not meet demand and the steps required to achieve the goal of the project's software products.

5. (　) Software engineering principle is to focus on engineering design, engineering

support, and must follow the principles of project management in the software development process.

6. (   ) In system design, software requirements, hardware requirements and other factors between the mutual restraint, mutual influence, and often need to weigh.

7. (   ) In software engineering, software tools and environment support for software process is quite important.

III. Translate the following words and phrases into Chinese.

1. Deployment View　　　　　　　　　　＿＿＿＿＿＿＿＿＿＿＿＿＿＿＿＿
2. Design Subsystem　　　　　　　　　　＿＿＿＿＿＿＿＿＿＿＿＿＿＿＿＿
3. Development Process　　　　　　　　＿＿＿＿＿＿＿＿＿＿＿＿＿＿＿＿
4. Double-byte Character Set(DBCS)　　＿＿＿＿＿＿＿＿＿＿＿＿＿＿＿＿
5. Executable Architecture　　　　　　　＿＿＿＿＿＿＿＿＿＿＿＿＿＿＿＿
6. Dynamic Link Library(DLL)　　　　　＿＿＿＿＿＿＿＿＿＿＿＿＿＿＿＿
7. Distributed Computing Environment (DCE)　＿＿＿＿＿＿＿＿＿＿＿＿＿＿＿＿
8. Software Development　　　　　　　　＿＿＿＿＿＿＿＿＿＿＿＿＿＿＿＿
9. Database Management System(DBMS)　＿＿＿＿＿＿＿＿＿＿＿＿＿＿＿＿
10. Disjoint Substrate　　　　　　　　　　＿＿＿＿＿＿＿＿＿＿＿＿＿＿＿＿

IV. Translate the following Chinese statements into English.

1. 系统建造这一工作(业务)已经变了。

2. 系统已经交到客户手中，并且正在处理真正的商业信息。

3. 通过注册安装一个新的系统而解决一个复杂的业务问题所获得的感觉是难以形容的。

4. 这些工业化实践活动在公司内升级的系统中得到具体化。

5. 业务分析和系统设计必须敏锐地把握围绕着系统实施或者改变的业务意识。

Ⅴ. Fill in each of the blanks with one of the words.

*in   therefore   software   recognize   the goal of   focus on*

Software engineering principle is to_____engineering design, engineering support, and must follow the principles of project management in the_____development process. Software engineering principles the following four software engineers' basic principles: Select the appropriate development paradigm of the principle and system design._____ system design, software requirements, hardware requirements and other factors between the mutual restraint, mutual influence, and often need to weigh._____, we must_____the variability of_____requirements definition, use of appropriate development paradigm to be controlled to ensure that software products meet user requirements. Use of appropriate design methods in software design, usually to consider the software's modular characteristics of abstraction and information hiding, localization, consistency and adaptability. Appropriate design approach helps to achieve these characteristics, in order to achieve _____software engineering.

## SUPPLEMENTARY

## About 4G

Each major evolution in mobile phone technology has been marked by a generation. Currently the third generation, better known as 3G, is taking the world by storm. However, many industry experts have predicted that 4G is just around the corner, so what is 4G?

### 1. What is 4G?

Mobile Internet still features as a major factor for innovation, as it still cannot hope to rival the PC-bound Internet experience. 4G is going to introduce wireless technologies to mainstream mobile phones.

Since 4G has not been officially defined by anyone, there is a lot of speculation as to what exactly will constitute 4G mobile technology. There are various contenders for the role; two of the frontrunners being WiMAX and LTE.

However, both the technologies have one common factor which seems to unanimously herald the onset of 4G: orthogonal frequency-division multiplexing or OFDM.

## 2. OFDM

Currently, there are a number of ways that data is transferred from device to device. CDMA (code division multiple access) is one popular technology, and TDMA (time division multiple access) is a second. CDMA transfers a number of different data packets on one channel, using codes to distinguish between different receivers. TDMA, on the other hand, again uses one channel but allots each different data packet a time slot.

OFDM differs from these technologies significantly; the channel itself is divided into narrow bands, and data packets are sent through each band individually. This method proves to be much more efficient than the previously used technologies, hence it is considered to be an integral part of the 4G revolution.

## 3. WiMAX

WiMAX, simply put, is broadband on a wireless connection. The existence of WiMAX will ensure that high-speed Internet access is available anywhere a user goes. The hardware is still in the design stages, because the adoption of this technology will entail a huge overhaul of existing telecommunication networks.

High-speed Internet access will drastically change a user's mobile Internet experience. As it stands, the user is restricted to certain web interactions like uploading photos or videos, and maybe updating a blog once in a while. The data transfer rates prevalent are exorbitant and therefore data transfer is severely limited. The implementation of high-speed data access on mobile phones will finally catapult the mobile phone into serious contention in the personal computer segment.

## 4. LTE: Long-Term Evolution

Developed by the 3GPP (Third Generation Partnership Project) group, LTE is a new wireless broadband technology that differs from WiMAX. LTE places great emphasis on IP addresses, since it is closely based upon the TCP/IP networking skeleton. The idea is to create enough IP addresses so that each device has a unique one.

However the main focus of LTE is that it is shaping up to be high performance in every way. The technology isn't fully developed as yet, but the characteristics that are emerging promise great things: high peak rates for data transmission, reduced latency and scalable bandwidth.

An important feature of LTE is the ability to co-exist harmoniously with previous architectures. Therefore a LTE-enabled network can pass data back and forth from a CDMA network seamlessly.

## 5. iPhone 4G

Hot on the heels of the iPad release, iPhone 4G handsets have been rumored for release this summer. Every year, since the first iPhone handset was launched, there has been a new version of the phone released regularly. This year promises to be no different.

Speculation is rife that the 4G tacked on the end of the new iPhone means that the handset may just be 4G enabled. The argument posed here is that the 3G model was actually the second phone, so it is probably not a version number as previously thought.

Apple hasn't confirmed or denied anything as yet; the new iPhone may not be 4G at all, but an HD (high definition) model. This scenario is certainly more credible as the improvements required for the iPhone to merit the 4G tag will be quite substantial.

## 6. 4G is MAGIC

4G networks are still in the fluid stages of development. As of now, it is sometimes referred to as MAGIC (Mobile multimedia, Anytime/any-where, Global mobility support, Integrated wireless and Customized personal service)—which bodes well for a mobile phone user. It will be interesting to watch the technologies unfold over the course of the next few months.

# CONVERSATION

## Telephone Marketing Skills

## 成功电话行销技巧

电话行销可不是拿起电话聊天就能算的,既然打电话的目的是约访,当然要有一些电话行销技巧来帮助你更快上手。

  技巧一，让自己处于微笑状态。微笑地说话，声音也会传递出很愉悦的感觉，客户听起来就会觉得有亲和力，让每一通电话都保持最佳的质感，并帮助你进入对方的时空。

  技巧二，音量与速度要协调。人与人见面时，都会有所谓"磁场"。在电话之中，当然也有电话磁场，一旦业务人员与客户的磁场吻合，谈起话来就顺畅多了。为了了解对方的电话磁场，建议在谈话之初，采取适中的音量与速度，等辨出对方的特质后，再调整自己的音量与速度，让客户觉得你和他是站在同一阵线上的。

  技巧三，判别通话者的形象，增进彼此互动。从对方的语调中，可以简单判别通话者的形象。讲话速度快的人是视觉型的人，说话速度中等的人是听觉型的人，而讲话慢的人是感觉型的人。业务人员可以在判别通话者的类型之后，再给对方"适当的建议"。

  技巧四，表明不会占用太多时间，简单说明。为了让对方愿意继续这通电话，最常用的方法就是请对方给两分钟，而一般人听到两分钟时，通常都会出现"反正才两分钟，就听听看好了"的想法。实际上，你真的只讲两分钟吗？这得看个人的功力了！

  技巧五，语气、语调要一致。在电话中，开场白通常是普通话，但是如果对方的反应是以方言回答，就要适当地用当地地方话和对方说话，有时用方言交谈也是一种拉近双方距离的方法，主要目的是为了要与对方处于同一个"磁场"。

  技巧六，善用电话开场白。好的开场白可以让对方愿意和业务人员多聊一聊，因此除了耽误两分钟之外，接下来该说些什么就变得十分重要，要想多了解对方的想法，不妨提一些开放式问句。

  技巧七，善用暂停与保留的技巧。什么是暂停？当业务人员需要对方给一段时间的时候，就可以使用暂停的技巧。比如，当你问对方问题时，说完后稍微暂停一下，让对方回答。善用暂停的技巧，将会让对方有受到尊重的感觉。

  至于保留，则是在业务人员不方便在电话中说明或者遇到难以回答的问题时所采用的方式。举例来说，当对方要求业务人员在电话中说明费率时，业务人员就可以告诉对方："这个问题在我们见面谈时再当面计算给您听，这样会比较清楚。"如此便可将问题保留到下一个时空，这也是约访时的技巧。

  技巧八，身体挺直、站着说话或闭上眼睛。假如一天打二十通电话，总不能一直坐着不动吧！试着将身体挺直或站着说话，你可以发现，声音会因此变得有活力，效果也会变得更好。有时不妨闭上眼睛讲话，让自己不被外在的环境影响谈话内容。

  技巧九，使用开放式问句，不断问问题。问客户问题，一方面可以拉长谈话时间，更重要的是了解客户真正的想法，帮助业务员做判断。

  技巧十，即时逆转。即时逆转就是立刻顺着客户的话走。

  技巧十一，一再强调由客户自己判断、客户自己做决定。为了让客户答应和你见面，在电话中强调由客户自己做决定，可以让客户感觉业务人员是有素质的、是不会纠缠不休的，进而提高约访概率。

  技巧十二，强调产品的功能或独特性。在谈话中，多强调产品很特别，再加上由客户自己做决定，让客户愿意将他宝贵的时间给你。切记千万不要说得太繁杂或使用太多专业术语，让客户失去见面的兴趣。

技巧十三，给予二选一的问题及机会。二选一方式能够帮助对方做选择，同时也可加快对方与业务人员见面的速度。

技巧十四，为下一次开场做准备。

# Example

## ✉ Dialogue 1

A: Madison Industries. This is Cathy Winner speaking. Can I help you?
B: Good afternoon. Could you connect this call with Mr. Black, please?
A: May I know whose calling?
B: This is Mary Fox of A.B.C. Computer Co. I'm calling on behalf of Mr. Tom Baker, the general manager of our company.
A: I am sorry, Ms. Fox. Mr. Black is now in a meeting. May I have your number and ask him to call back later?
B: I'm afraid Mr. Baker would like to speak to Mr. Black right now. He has got an urgent matter to discuss with Mr. Black without delay.
A: OK. Then, would you please hold the line? (One minute later.)
   Ms. Fox, the line is through. Mr. Black is ready to answer the call. Go ahead.
B: Thank you for your kind assistance, Ms. Winner.
A: You are welcome.

## ✉ Dialogue 2

Office Assistant: Good morning. Odyssey Promotions. How may I help you?
Nick: Hello, this is Nick Delwin from Communicon. Could I speak to Helen Turner, please?
Office Assistant: Just a moment, please.
Office Assistant: I have Nick Delwin on the line for you.
Helen: Thank you… Hi, Nick. Nice to hear from you. How's the weather?
Nick: It's pretty good for the time of year. What's it like in New York?
Helen: Not good, I'm afraid.
Nick: That's a pity because I'm planning to come across next week.
Helen: Really? Well, you'll come by to see us while you're here, I hope?
Nick: That's what I'm phoning about. I've got a meeting with a customer in Boston on Tuesday of next week. I was hoping we could arrange to meet up either before or after.

| | |
|---|---|
| Helen: | Great. That would give us a chance to show you the convention centre, and we could also drop in at Caesar's Restaurant where Gregg has arranged your reception. |
| Nick: | That's what I was thinking. |
| Helen: | So you said you have to be in Boston on Tuesday? That's the 8th? |
| Nick: | That's right. Now, I could stop over in New York either on the way in—that would be the Monday…Would that be possible? |
| Helen: | Ah, I'm afraid I won't be in the office on Monday, and I think Gregg has meetings all day. |
| Nick: | Uh-huh, well, the other possibility would be to arrange it after Boston on my way home. |
| Helen: | When do you plan on leaving Boston? |
| Nick: | Could be either Tuesday afternoon or Wednesday morning, but I would like to catch a flight back to London on Wednesday evening. |
| Helen: | OK. Well, it would be best for us if you could fly in on the Wednesday morning. Either Gregg or I will pick you up at the airport, and then we could show you the convention centre and also Caesar's. If there's time, you could come back to the office and we'll run through any of the details that still haven't been finalized. |
| Nick: | That sounds good. Just as long as I can get back to the airport for my evening flight. |
| Helen: | No problem. Look, why don't you fax me your information once you've confirmed your flight times? Then we'll get back to you with an itinerary for the day—that's Wednesday the 9th, right? |
| Nick: | That's right. Good, well, I'll do that and I look forward to seeing you next week. |
| Helen: | Same here. See you next week. |
| Nick: | Right. Goodbye. |
| Helen: | Bye-bye. |

## Practice

Imagine you are the boss of AAA company, you want to discuss business with the boss of BBB company. Practice telephone booking skills with the given points:

- ◇ Greet each other.
- ◇ Outline the current problems they are facing.
- ◇ Look at the causes of the damage.
- ◇ Find out the ways to solve these problems.
- ◇ Discuss how to carry out the contract smoothly.

# Tips

## ✉ 电话预约常用语

1. Is Daisy there? This is she/he. 黛西在吗？这是他/她。

2. May I speak to Mr. Gates? 我可以和盖斯先生说话吗？

3. He's not here right now. 他现在不在这里。

4. He's in a meeting right now. 他现在正在开会。

5. You've just missed him. 你刚好错过他了。

6. He's just stepped out. 他刚好出去了。

7. He's out on his lunch break right now. Would you like to leave a message? 他出去吃午饭了，你要留言吗？

8. He's not available right now. Can I take a message? 他不在，我可以帮你传话吗？

9. Do you know when he will be back? 你知道他什么时候会回来吗？

10. I have no idea. 我不知道。

11. He should be back in 20 minutes. 他应该在二十分钟内会回来。

12. Do you have any idea where he is? 你知道他在哪里吗？

13. He's at work right now. Do you want his phone number? 他现在在上班。你要不要他的电话号码？

14. Can I leave a message? 我可以留个话吗？

15. Yes. Go ahead, please. 可以，请继续。

16. Of course. Hold on for just a second so I can grab a pen and paper. 当然，稍等一下，让我去拿纸和笔。

17. When he comes back, can you have him call me at 123456789? 他回来后请让他打 123456789 这个号码给我。

18. Can you repeat again, please? 能否重复一遍？

19. (Say) Again, please? 再说一次好吗？

20. Pardon? 请再说一遍。

 WRITING

# How to Write IOU and Receipt in English

# 如何写借据与收据

日常生活中常常发生物品和金钱的借进、借出或转手,这些时候往往需要写借据(IOU,即 I owe you,译成中文为"今欠"、"今借到")或收据(Receipt),这既是手续,也是必要的证据。收据种类很多,有收条、订阅单、订货单等,是在跟对方发生钱和物的关系时写给对方作为凭据的条子,起书面证据作用。

借据和收据一般篇幅短小,内容主要包括双方当事人的全名,具体物款的名目、数额以及收到、借出和归还的准确日期。借据和收据的格式类似于便条,基本句型比较固定:借据的表述一般是 "IOU+具体款数(sum of money)+only",收据是 "Received from sb. +具体款数(sum of money)"。借据和收据的内容均有正式和非正式之分。正式借据要写明归还日期。在写借据、收条时,写字据的日期写于右上角,立据人写于右下角。正式的收据常用 "for/being something for..." 说明缘由或物品、款项的用途。

注意事项:在正式的英文借据和收据中,归还日期一般不用阿拉伯数字,而用英文单词表示。物款的数额一般要同时用阿拉伯数字和英文单词表示。

## Example

May 8, 2012

To Mr. Charles Green,

IOU three thousand U.S. dollars(US $3000)only, within one year from this date with annual interest at four percent(4%).

David Smith

✉ 给查尔斯·格林先生：

兹借查尔斯·格林先生叁仟美元(US $3000)，年息四厘，自即日算起，一年内归还。

借款人：戴维·史密斯

2012年5月8日

---

June 8, 2012

Received from Mr. Handel the following things:
One typewriter
One tape-recorder

Bruce

---

兹收到汉德尔先生下述物件：打字机壹台，录音机壹台。

布鲁斯

2012年6月8日

## Practice

To Mr. Smith,

_____

_____

_____

_____

✉ 给史密斯先生：

兹借人民币2000元整，自即日算起，两星期内归还。

借款人： 李四

May 8, 2012

今收到何小姐的还款1000元人民币。

张三

2012年5月8日

## Tips

### 几种常见的借据

Oct.21, 2011

To Ms. Nancy Wells,

　　IOU　two hundred US dollars(US $200) only.

Mary Brown

Nanjing, July 12, 2011

US $300

Eight months after date for value received, we jointly promise to pay Miss Helen White the sum of three thousand US dollars with interest at 2% per annum.

T. Brown

J. Robinson

D. White

☒                                                                Aug. 13, 2011

One IBM Laptop

Three weeks after date I promise to return the computer to Mr. Zhang Xiaojun from the Dept. of Computer Science.

Liang Qun

The History Department

# UNIT 5

## TOPICS

- What is database?
- What is the DBMS?
- What is the database model?
- What is the SQL? How much do you know about NoSQL?
- What is the function of SQL?
- What is the main future of database?
- What is composed of the DBMS?
- What is the distributed database?
- What are the language elements that SQL language sub-divides into?

# An Introduction to SQL

## Introduction

Structured Query Language, commonly **abbreviated** to SQL and pronounced as sequel, is not a **conventional** computer programming language in the normal sense of the phrase. With SQL you don't write applications, utilities, batch processes, GUI interfaces or any of the other types of program for which you'd use languages such as Visual Basic, C++, Java etc. Instead SQL is a language used exclusively to create, **manipulate** and **interrogate** databases.

SQL is about data and results, each SQL statement returns a result, whether that result be a **query**, an update to a record or the creation of a database table. A **query optimizer** translates the description into a **procedure** to perform the database manipulation.

SQL was one of the first **commercial** languages for Edgar F. Codd's relational model, as described in his **influential** 1970 **paper**, "A Relational Model of Data for Large Shared Data Banks". **Despite** not adhering to the relational model as described by Codd, it became the most widely used database language.

## Relational Databases

Before looking at SQL itself it's worth looking more closely at what is meant by a database. Although some people talk about SQL databases, there is in fact no such

abbreviate [ə'briːvieɪt] vt. 缩写；缩短；使简略
conventional [kən'venʃənl] adj. 传统的；惯例的；常规的

manipulate [mə'nɪpjuleɪt] vt. 操纵；操作；控制；利用
interrogate [ɪn'terəgeɪt] v. 质问；讯问；审问
query ['kwɪəri] n. 疑问；质问；疑问号　vt. 质问；对……表示怀疑
query optimizer 查询优化器
procedure [prə'siːdʒə(r)] n. 程序；手续；步骤
commercial [kə'məːʃəl] adj. 商业的；营利的
influential [ˌɪnflu'enʃəl] adj. 有影响的；有势力的
paper ['peɪpə] n. 纸；论文；文件；报纸
despite [dɪ'spaɪt] prep. 尽管；不管　n. 憎恨；轻视

thing. SQL itself makes no references to the underlying databases which it can access, which means that it is possible to have a SQL **engine** which can **address** relational databases, non-relational (flat) databases, even **spreadsheets**. However, SQL is most often used to address a relational database, which is what some people refer to as a SQL database.

engine ['endʒɪn] n. 发动机；引擎；机车；火车头 vt. 给……装引擎

address [ə'dres] n. 住址；网址；称呼；致词；讲话；演讲；谈吐
v. 称呼；发表演说；提出；写地址；处理

spreadsheets ['spredʃiːt] n. 电子制表软件；电子表格；试算表

The relational model of a database was first defined by Dr E.F Codd in 1970, working at the IBM Research Labs at San Jose, and it describes a way of structuring data which is **mathematically** consistent and **abstracted** away from the physical **implementation** of the database.

The key concept of the relational model is that data sits in tables, and that data elements within different tables can be related in some way to provide meaningful information rather than just lists of data. Each table in the database has a number of **fields** which define the content and a number of **instances** of that field—think of the fields as **columns** in a table and the instances as rows. For example if we had a table of LAN users, then it might have fields of first name, surname and userID. The rows in such a table would contain the values which define each instance of user, for example the entry to define user John Smith might be: John, Smith, JSmith.

mathematically [ˌmæθə'mætɪkli] adv. 精确地；以数学

abstract ['æbstrækt] adj. 抽象的；理论的；抽象派的 v. 把……抽象出；提取；抽取

implementation [ˌɪmplɪmen'teɪʃn] n. 履行；落实；装置

fields ['fiːldz] n. 域；字段

instance ['ɪnstəns] n. 例子；实例；情况；要求

column ['kɒləm] n. 栏；专栏；列；圆柱；柱；柱形物

## A Word About Datatypes

Many kinds of data can be stored in a database, from simple text to various types of number to **Boolean** flags to binary objects and graphics. This is one area where there are differences between different products. Not just in the formats of data that can be stored, but also in how these formats are named. Boolean fields, for example, are called Yes/No fields in Microsoft Access, and BOOLEAN in the later releases of MySQL.

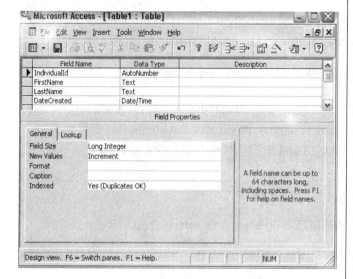

## Creating a Database

SQL commands follow a number of basic rules. SQL keywords are not normally **case sensitive**, though in this passage all commands (SELECT, UPDATE etc.) are upper-cased. Variable and parameter names are displayed here as lower-case. New-line characters are **ignored** in SQL, so a command may be all on one line or broken up across a number of lines **for the sake of** clarity. Many DBMS systems expect to have SQL commands terminated with a **semi-colon** character. Finally, although there is a SQL standard, actual implementations vary by **vendor**/system, so if in doubt always refer to the documentation that comes with your DBMS.

datatype 数据类型

Boolean ['buːliən] n. [数学]布尔 adj. [数学]布尔逻辑的

case sensitive 区分大小写

ignore [ɪgˈnɔː(r)] vt. 忽视；不理；不顾

for the sake of 出于……的缘故，为了……的利益

clarity [ˈklærəti] n. 清楚；透明

semi-colon [ˈsemɪˌkəʊlɒn] n. 分号

vendor [ˈvendə(r)] n. 自动售货机；小贩；卖方；供货商
=vender

# EXERCISES

I. Match the terms and interpretations.

1. SQL
2. Relational Database
3. Boolean
4. Database
5. RDBMS

(a) A collection of data arranged for ease and speed of search and retrieval. Also called data bank.

(b) Structured Query Language, a database computer language designed for managing data in relational database management systems.

(c) A database in which relations between information items are explicitly specified as accessible attributes.

(d) Relational database management systems.

(e) Of or relating to a combinatorial system devised by George Boole that combines propositions with the logical operators AND and OR and IF THEN and EXCEPT and NOT.

II. Are the following statements True (T) or False (F)?

1. (　) SQL is case sensitive, so SELECT is different from select.
2. (　) A query optimizer translates the description into a procedure to perform the database manipulation.
3. (　) SQL itself makes no references to the underlying databases which it can access.
4. (　) SQL can access and manipulate data in various database systems.
5. (　) The same data stored in the same format in different database systems.
6. (　) New-line characters are ignored in SQL.
7. (　) Many DBMS systems expect to have SQL commands ended with a period character.
8. (　) SQL is the standard language for data manipulation, so it implemented the same way on various DBMS.
9. (　) Each table in the database has a number of rows but only one field.
10. (　) The rows in a table would contain the values which define each instance of an entity set.

III. Translate the following words and phrases into Chinese.

1. SQL _____

2. Query Optimizer _____

3. DBMS _____

4. Data Manipulation _____

5. Data Types _____

6. Case Sensitive _____


IV. Translate the following Chinese statements into English.

1. 数据库管理系统相当于用户、计算机操作系统和数据库之间的接口。
_____

2. 数据库是存储在一起的相关数据的集合。
_____

3. 多个用户可以同时共享数据库中的数据资源，即不同的用户可以同时读取数据库中的同一个数据。
_____
_____

4. 在关系数据库中，对数据的操作几乎全部建立在一个或多个关系表格上。
_____

5. SQL 语言的功能包括查询、操作、定义和控制，是一个综合的、通用的关系数据库语言，同时又是一种高度非过程化的语言。
_____
_____


V. Fill in each of the blanks with one of the following words or phrases.

*relational    classified    a collection of    according to    accessed*

*standard    distributed    object-oriented    typically    catalogs*

A database is_____information that is organized so that it can easily be_____, managed, and updated. In one view, databases can be classified_____types of content: bibliographic, full-text, numeric, and images.

In computing, databases are sometimes_____according to their organizational approach. The most prevalent approach is the_____database, a tabular database in which data is defined so that it can be reorganized and accessed in a number of different ways. A_____database is one that can be dispersed or replicated among different points in a network. An_____programming database is one that is congruent with the data defined in object classes and subclasses.

Computer databases_____contain aggregations of data records or files, such as sales transactions, product_____and inventories, and customer profiles. Typically, a database manager provides users the capabilities of controlling read/write access, specifying report generation, and analyzing usage. Databases and database managers are prevalent in large mainframe systems, but are also present in smaller distributed workstation and mid-range systems such as the AS/400 and on personal computers. SQL (Structured Query Language) is a _____language for making interactive queries from and updating a database such as IBM's DB2, Microsoft's SQL Server, and database products from Oracle, Sybase, and Computer Associates.

## SUPPLEMENTARY

## Database Management System and Database Model

A database (sometimes spelled data base) is also called an electronic database, referring to any collection of data or information that is specially organized for rapid search and retrieval by a computer. Databases are structured to facilitate the storage, retrieval, modification, and deletion of data in conjunction with various data-processing operations. Databases can be stored on magnetic disk or tape, optical disk, or some other secondary storage device.

A database consists of a file or a set of files. The information in these files may be broken down into records, each of which consists of one or more fields. Fields are the basic units of data storage, and each field typically contains information pertaining to one aspect or attribute of the entity described by the database. Using keywords and various sorting commands, users can rapidly search, rearrange, group, and select the fields in many records to retrieve or create reports on particular aggregate of data.

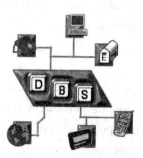

Complex data relationships and linkages may be found in all but the simplest databases. The system software package that handles the difficult tasks associated with creating, accessing, and maintaining database records is called a database management system(DBMS). The programs in a DBMS package establish an interface between the database itself and the users of the database. (These users may be applications programmers, managers and others with information needs, and various OS programs.)

A DBMS can organize, process, and present selected data elements form the database. This capability enables decision makers to search, probe, and query database contents in order to extract answers to nonrecurring and unplanned questions that aren't available in regular reports. These questions might initially be vague and/or poorly defined, but people can "browse" through the database until they have the needed information. In short, the DBMS will "manage" the stored data items and assemble the needed items from the common database in response to the queries of those who aren't programmers.

A database management system (DBMS) is composed of three major parts:

(1) a storage subsystem that stores and retrieves data in files;

(2) a modeling and manipulation subsystem that provides the means with which to organize the data and to add, delete, maintain, and update the data;

(3) and an interface between the DBMS and its users.

Several major trends are emerging that enhance the value and usefulness of database management systems:

(1) Managers require more up-to-data information to make effective decision.

(2) Customers demand increasingly sophisticated information services and more current

information about the status of their orders, invoices, and accounts.

(3) Users find that they can develop custom applications with database systems in a fraction of the time it takes to use traditional programming languages.

(4) Organizations discover that information has a strategic value; they utilize their database systems to gain an edge over their competitors.

**The Database Model**

A database model describes a way to structure and manipulate the data in a database. The structural part of the model specifies how data should be represented(such as tree, tables, and so on). The manipulative part of the model specifies the operation with which to add, delete, display, maintain, print, search, select, sort and update the data. The commonly classification of the database model:

**1. Hierarchical Model**

The first database management systems used a hierarchical model-that is-they arranged records into a tree structure. Some records are root records and all others have unique parent records. The structure of the tree is designed to reflect the order in which the data will be used that is, the record at the root of a tree will be accessed first, then records one level below the root, and so on.

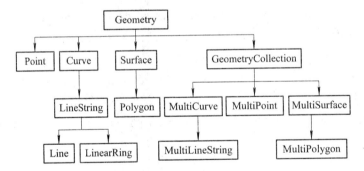

The hierarchical model was developed because hierarchical relationships are commonly found in business applications. As you have known, an organization chart often describes a hierarchical relationship: top management is at the highest level, middle management at lower levels, and operational employees at the lowest levels. Note that within a strict hierarchy, each level of management may have many employees or levels of employees beneath it, but each employee has only one manager. Hierarchical data are characterized by this one-to-many relationship among data.

In the hierarchical approach, each relationship must be explicitly defined when the database is created. Each record in a hierarchical database can contain only one key field and only one relationship is allowed between any two fields. This can create a problem because data do not always conform to such a strict hierarchy.

## 2. Relational Model

A major breakthrough in database research occurred in 1970 when E. F. Codd proposed a fundamentally different approach to database management called relational model, which uses a table as its data structure.

The relational database is the most widely used database structure. Data is organized into related tables. Each table is made up of rows called records and columns called fields. Each record contains fields of data about some specific item. For example, in a table containing information on employees, a record would contain fields of data such as a person's last name, first name, and street address.

Relational Model

| Activity Code | Activity Name |
|---|---|
| 23 | Patching |
| 24 | Overlay |
| 25 | Crack Sealing |

Key=24

| Activity Code | Date | Route No. |
|---|---|---|
| 24 | 01/12/01 | I-95 |
| 24 | 02/08/01 | I-66 |

| Date | Activity Code | Route No. |
|---|---|---|
| 01/12/01 | 24 | I-95 |
| 01/15/01 | 23 | I-495 |
| 02/08/01 | 24 | I-66 |

## 3. Network Model

The network model creates relationships among data through a linked list structure in which subordinate records can be linked to more than one parent record. This approach combines records with links, which are called pointers. The pointers are addresses that indicate the location of a record. With the network approach, a subordinate record can be linked to a key record and at the same time itself be a key record linked to other sets of subordinate records. The network mode historically has had a performance advantage over other database models. Today, such performance characteristics are only important in high-volume, high-speed transaction processing such as automatic

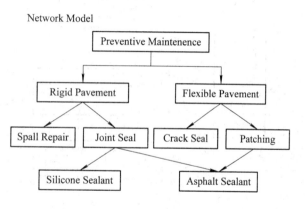

Network Model

teller machine networks or airline reservation system.

Both hierarchical and network databases are application specific. If a new application is developed, maintaining the consistency of databases in different applications can be very difficult. For example, suppose a new pension application is developed. The data are the same, but a new database must be created.

## 4. Object Model

The newest approach to database management uses an object model, in which records are represented by entities called objects that can both store data and provide methods or procedures to perform specific tasks.

The query language used for the object model is the same object-oriented programming language used to develop the database application. This can create problems because there is no simple, uniform query language such as SQL. The object model is relatively new, and only a few examples of object-oriented database exist. It has attracted attention because developers who choose an object-oriented programming language want a database based on an object-oriented model.

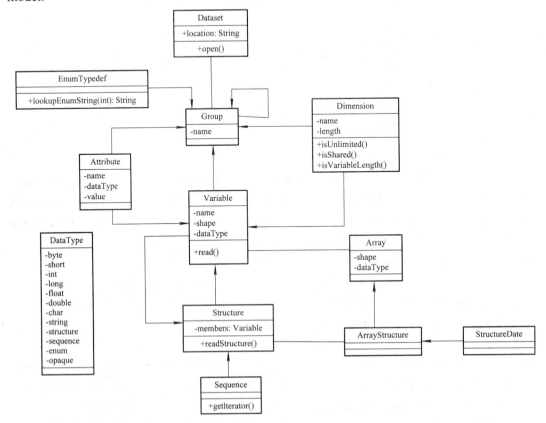

# CONVERSATION

# Booking Hotels

# 预 订 酒 店

### 1. 酒店预订的方式

酒店预订的方式通常有以下几种：
- 电话预订(Telephone);
- 传真订房(FAX);
- 国际互联网预订(Internet);
- 信函订房(Mail);
- 口头订房(Verbal);
- 合同订房(Contract)，指酒店与旅行社或商务公司之间通过签订订房合同，达到长期出租客房的目的。

### 2. 酒店预订的种类

(1) 临时预订(Advance Reservation)：客人在即将抵达酒店前很短的时间内或在到达的当天联系订房。

(2) 确认类预订(Confirmed Reservation)：酒店可以事先声明为客人保留客房至某一具体时间，过了规定时间，客人如未抵达酒店，也未与酒店联系，则酒店有权将客房出租给其他客人。

(3) 保证类预订(Guaranteed Reservation)：指客人保证前来住宿，否则将承担经济责任，因而酒店在任何情况下都应保证落实的预订。

保证类预订又分为三种类型：
- 预付款担保;
- 信用卡担保;
- 合同担保。

### 3. 预订渠道

预订渠道通常有以下几种：
- 散客自订房，可以通过电话、互联网、传真等方式进行;
- 旅行社订房;
- 公司订房;
- 各种国内外会议组织订房;

- 分时度假(timeshare)组织订房;
- 国际订房网络组织订房;
- 其他组织订房。

4. 国际酒店通行的几种收费方式

(1) 欧洲式(European Plan, EP): 只包括房费而不包含任何餐费的收费方式,为世界上大多数酒店所采用。

(2) 美国式(American Plan, AP): 不但包括房费,而且还包括一日三餐的费用,因此又被称为"全费用计价方式",多为远离城市的度假性酒店或团队客人所采用。

(3) 修正美式(Modified American Plan, MAP): 包括房费和早餐,除此而外,还包括一顿午餐或晚餐(二者任选其一)的费用。这种收费方式较适合于普通旅游客人。

(4) 欧洲大陆式(Continental Plan, CP): 包括房费及欧洲式早餐(Continental Breakfast)。欧陆式早餐的主要内容包括冷冻果汁(Orange Juice、Grape Juice、Pineapple Juiciest 等)、烤面包(Served with Butter &Jam)、咖啡或茶。

(5) 百慕大式(Bermuda Plan, BP): 包括房费及美式早餐(American Breakfast)。美式早餐除了包含欧陆式早餐的内容以外,通常还包括鸡蛋(Fried、Scrambled up、Poached、Boiled)、火腿(Ham)、香肠(Sausage)或咸肉(Bacon)等。

# Example

## Dialogue 1

F: Hello, welcome to Prise Star Hotel. How may I do for you?

M: Hi, yes, I have a reservation. My secretary called and booked a room a couple of weeks ago; the reservation should be for a double bed, none smoking room.

F: And what name of reservation made under?

M: It should be under Steve Johnson.

F: Mmm, let me see, it seems there is no Johnson listed for a room for tonight, is there any other name that you reservation list under?

M: No, here is the confirmation number, would that help? It is 898007, I had the room booked with a visa cord card.

F: Ah, yes, here it is, you have a standard double room, non smoking on the 3rd floor, I just need to see some identification and the credit card you booked the room with if you don't mind.

M: Sure, here it is. Would it be possible to check out and pay the bill in the morning, also, what time is breakfast served?

F: There is a continental breakfast buffet from 6:00am to 10:00am, it's in the lobby.

M: Thank you!

F: You're welcome!

## ✉ Dialogue 2

> A: Hello, this is QuanCheng Hotel, is there anything I can help you, sir?
> B: Yes, I'd like to book a room.
> A: Sure, we have different type of rooms, which would you like to choose?
> B: I'm not very certain; could you give me a brief introduction, please?
> A: OK, our hotel provides standard single room and double room; accordingly, we also have special treatment for VIP.
> B: Thank you, I want to reserve a single room which is better located between 4th floor to 7th floor. Well, I need the personnel in your hotel to open the window and clean the room before I arrive there. Is that OK?
> A: OK, sir, I already taken some notes about your requirements. Now let me check the room on my computer. Well, one standard single room in 6th floor and the number is 0603, need clean and open the window, anything else?
> B: Thanks, how much does it cost?
> A: That is 580 RMB.
> B: Don't your hotel have discount?
> A: I'm terribly sorry, sir. If you need a discount, you should be our VIP and stay here for at least 3 days.
> B: OK, that's alright!

## Practice

Imagine that Mary is the Front Desk of the ABC Hotel, and John Deep is going to reserve a room from 2nd to 4th, May. John Deep is calling to Mary. Please complete the following dialogue between Mary and John with the given points:

- ✧ Greet each other.
- ✧ Describing the needed room that should be a suite with an ocean view.
- ✧ Talking the discount and offer ended yesterday.
- ✧ Talking about the breakfast and the check-in time.

## Tips

### ✉ 酒店预订常用语

1. Room Reservations. May I help you? 房间预订，有什么能帮助您的吗？
2. I'd like to book a double room for Tuesday next week. 下周二我想订一个双人房间。

3. What's the price difference? 两种房间的价格有什么不同？

4. A double room with a front view is 140 dollars per night, one with a rear view is 115 dollars per night. 一间双人房朝阳面的每晚 140 美元，背阴面的每晚 115 美元。

5. I think I'll take the one with a front view then. 我想我还是要阳面的吧。

6. How long will you be staying? 您打算住多久？

7. We'll be leaving Sunday morning. 我们将在星期天上午离开。

8. And we look forward to seeing you next Tuesday. 我们盼望下周二见到您。

9. I'd like to book a single room with bath from the afternoon of October 4 to the morning of October 10. 我想订一个带洗澡间的单人房间，10 月 4 日下午到 10 月 10 日上午用。

10. We do have a single room available for those dates. 我们确实有一个单间，在这段时间可以用。

11. What is the rate, please? 请问房费多少？

12. The current rate is $50 per night. 现行房费是 50 美元一天。

13. What services come with that? 这个价格包括哪些服务项目呢？

14. That sounds not bad at all. I'll take it. 听起来还不错。这个房间我要了。

15. By the way, I'd like a quiet room away from the street if the is possible. 顺便说一下，如有可能我想要一个不临街的安静房间。

16. I'd like to confirm my reservation. This is XXX from Hong Kong. I've reserved a double room for January second. 我想确定我的订房结果。我是从香港来的 XXX。我已经预订了一间供 1 月 2 日入住的双人房。

17. May I order my breakfast for tomorrow morning? 我可以预订明天早上的早餐吗？

18. By the way, you might have my things sent up here. 顺便说一句，请您把我的东西搬到这儿来。

19. I'd like to cancel my reservation and make a new reservation. 我想取消我原来的预订，并作新的预订。

20. This is rather expensive. I must have a cheaper one. Can't you show me something less expensive? 这有点贵。我要一间比较便宜的。可否给我看些较廉价的房间？

21. Yes, we do have a reservation for you. 对了，我们这儿是有您预订的房间。

22. Would you please fill out this form while I prepare your key card for you? 请您把这份表填好，我同时就给您开出入证，好吗？

23. What should I fill in under ROOM NUMBER? "房间号码"这一栏我该怎么填呢？

24. I'll put in the room number for you later on. 过会儿我来给您填上房间号码。

25. Please make sure that you have it with you all the time. 请务必随时带着它。

26. And here is your key, Mr. Bradley. Your room number is 1420. 给您房间的钥匙，布拉德利先生。您的房间号码是1420。

27. It is on the 14th floor and the daily rate is $90. 房间在14层，每天的房费是90美元。

28. I wonder if it is possible for me to extend my stay at this hotel for two days. 我想知道是否可以让我在这儿多待两天。

29. I'll take a look at the hotel's booking situation. 我来查看一下本店房间的预订情况。

30. This is a receipt for paying in advance. Please keep it. 这是预付款收据，请收好。

 WRITING

# How to Write Business Contract

# 如何写商务合同

1. Ask your client to list the deal points. This can be in the form of a list, outline or narration. Doing this will help the client focus on the terms of the agreement.

   要求你的客户列出合同交易的要点，也可以说是合同的清单、目录或概述。这一招首先帮助你的客户弄清合同的重点所在。

2. Engage your client in "what if" scenarios. A good contract will anticipate many possible factual situations and express the parties' understanding in case those facts arise. Talking to your client about this will generate many issues you may not otherwise consider.

   让你的客户提供一些假设可能发生的情况。好的合同不仅能够预见到许多可能发生的情况，而且还能清楚地描述出发生这些情况后合同双方的立场。和客户聊这些情况将有助于你发现一些你可能没有考虑到的问题。

3. Search your office computer or the Internet for a similar form. Many times you can find a similar form on your computer. It may be one you prepared for another client or one you negotiated with another lawyer. Just remember to find and replace the old client's name. Starting with an existing form saves time and avoids the errors of typing.

   在办公室的电脑中或是在因特网上搜索类似的合同范本。通常你会在你的电脑上找到你想要的东西，这些类似的合同范本要么是你给其他客户准备的，要么是你和其他的律师共同协商起草的。使用这些旧合同可以为你节省时间和避免打印错误。

4. Don't let your client sign a letter of intent without this wording. Sometimes clients are anxious to sign something to show good faith before the contract is prepared. A properly

worded letter of intent is useful at such times. Just be sure that the letter of intent clearly states that it is not a contract, but that it is merely an outline of possible terms for discussion purposes.

如果没有特别申明,不要让你的客户在意向书上签字。有时候,在合同未准备好之前,客户为了表示诚意,往往急于签署某些东西。当然,在这种情况下,如果客户急于签署的是有特别申明的意向书,这也是可以的,但一定要注意:本意向书并非合同,只是双方为了更好地沟通协商而拟定的对未来条款的概述。

5. State the correct legal names of the parties in the first paragraph. As obvious as this is, it is one of the most common problems in contracts. For individuals, include full first and last name, and middle initials if available, and other identifying information, if appropriate, such as Jr., M.D., etc. For corporations, check with the Secretary of State where incorporated.

在合同的第一段要写清楚双方的名称,这是个简单而又不得不引起重视的问题。如果是个人,要写清姓和名,中间有大写字母和其他身份信息的,也要注明,例如 Jr., M.D 等;如果是公司,为避免弄错,写名称时可以到公司注册地的相应机构去核对一下。

6. Be careful when using legal terms for nicknames. Do not use "Contractor" as a nickname unless that party is legally a contractor. Do not use "Agent" unless you intend for that party to be an agent, and if you do, then you better specify the scope of authority and other agency issues to avoid future disagreements.

使用法定术语作为双方当事人的别称时要小心。除非一方当事人在法定上就是承包人,否则不要将"承包人"作为其别称。同样,除非你想让一方当事人成为法定上的代理人,否则不要称其为"代理人",如果坚持要用,最好明确一下代理范围并找到其他可以避免将来争执的方案。

7. Watch out when using "herein". Does "wherever used herein" mean anywhere in the contract or anywhere in the paragraph? Clarify this ambiguity if it matters.

使用术语"本文"(herein,也可译为"在这里")时要当心。为了避免含糊不清,使用"本文"时最好特别申明一下"本文"是指整个合同,还是指其所在的某一段落。

8. Write numbers as both words and numerals: ten (10). This will reduce the chance for errors.

写数目时要汉字和阿拉伯数字并用,如拾(10)。这将减少一些不经意的错误。

9. When you write "including" consider adding "but not limited to." Unless you intend the list to be all-inclusive, you had better clarify your intent that it is merely an example.

如果你想用"包括"这个词,就要考虑在其后加上"但不限于……"的分句。除非你能够列出所有被包括的项,否则最好用"但不限于……"的分句,来说明你只是想举个例子。

10. Be consistent in using words. If you refer to the subject matter of a sales contract as "goods" use that term throughout the contract; do not alternately call them "goods" and "items". Maintaining consistency is more important than avoiding repetition. Don't worry about putting the reader to sleep; worry about the opposing lawyer a year from now hunting for ambiguities to get your contract into court.

用词一致。在一份销售合同中，如果你想用"货物"来指整个合同的标的物，就不要时而称它们为"货物"，时而又改称它们为"产品"。保持用词一致性比避免重复更加重要。不要担心这会让读者打瞌睡，你应该提防的是对方律师会因为含糊不清的合同而将你告上法庭。

## Example

Should either of the parties to the contract be prevented from executing the contract by force majeure, such as earthquake, typhoon, flood, fire, war or other unforeseen events, and their occurrence and consequences are unpreventable and unavoidable, the prevented party shall notify the other party by telegram without any delay, and within 15 days thereafter provide detailed information of the events and a valid document for evidence issued by the relevant public notary organization explaining the reason of its inability to execute or delay the execution of all or part of the contract.

在合作期间，由于地震、台风、水灾、火灾、战争或其他不可预见并且对其发生和后果不能防止及避免的不可抗力事故，致使直接影响合同的履行或者不能按约定的条件履行时，遇有上述不可抗力的一方，应立即将事故情况电报通知对方，并应在15天内提供事故的详细情况及合同不能履行，或者部分不能履行，或者需要延期履行的理由的有效证明文件。此项证明文件应由事故发生地区的有权证明的机构出具。

This contract is signed by the authorized representatives of both parties on Dec. 9, 1999. After signing the contract, both parties shall apply to their respective Government Authorities for ratification. The date of ratification last obtained shall be taken as the effective date of the contract. Both parties shall exert their utmost efforts to obtain the ratification within 60 days and shall advise the other party by telex and thereafter send a registered letter for confirmation.

本合同由双方代表于1999年12月9日签订。合同签订后，由各方分别向本国政府当局申请批准，以最后一方的批准日期为本合同的生效日期，双方应力争在60天内获得批准，用电传通知对方，并用挂号信件确认。

# Practice

To: DHL-Sinotrans Customer Service Department

I have been well advised that _____
_____
_____,
I hereby authorize DHL-Sinotrans to process the following special handling (please tick at the service type you choose) on the shipment via DHL airway bill No. _____ (Please fill in the AWB number) and agree to pay for any extra charge e.g. freight, duty & tax and so on caused by this arrangement.
_____

Yours faithfully,

ABC Co.,Ltd

☒ 确认函

致：中外运-敦豪公司客户服务部

我司已知悉此类特殊安排存在风险并且 DHL 无法就成功完成此类特殊安排作出承诺。现我司确认中外运敦豪公司对快件--------------(请填写运单号码)进行如下特殊安排(请勾选以下服务类别选项)，并同意支付由此产生的运费、海关税费或其他附加费用。对于未成功操作的服务，DHL 将不收取服务费。　　　　　　　　　　　　ABC 有限公司

Article 18 Miscellaneous

1. The penalty, compensations, storage expenses and various economic losses as specified in this contract shall be paid within_ _ _ _ _ _days after the confirmation of the responsibilities by the settlement method_____
_____

2. The penalty specified in this contract is deemed as the compensations for losses. Where _____
_____
_____,
including the benefit after the implementation of this contract, but not exceeding the predicted losses caused by the breach of this contract when the party in breach signs this contract.

- 第十八条 其他事项
    1. 按本合同规定应付的违约金、赔偿金、保管保养费和各种经济损失，应当在明确责任后_____日内，按银行规定的结算办法付清，否则按逾期付款处理。
    2. 约定的违约金，视为违约的损失赔偿。双方没有约定违约金或预先赔偿额的计算方法的，损失赔偿额应当相当于违约所造成的损失，包括合同履行后可获得的利益，但不得超过违反合同一方订立合同时应当预见到的因违反合同可能造成的损失。

# Tips

## 商务合同常用语

1. The contract is made out in English and Chinese languages in quadruplicate, both texts being equally authentic, and each Party shall hold two copies of each text.

    本合同用英文和中文两种文字写成，一式四份。双方执英文文本和中文文本各一式两份，两种文字具有同等效力。

2. The contract shall be valid for 10 years from the effective date of the contract, on the expiry of the validity term of contract, the contract shall automatically become null and void.

    本合同有效期从合同生效之日算起共 10 年，有效期满后，本合同自动失效。

3. The parties hereto shall, first of all, settle any dispute arising from or in connection with the contact by friendly negotiations.

    双方首先应通过友好协商，解决因合同而发生的或与合同有关的争议。

4. The outstanding claims and liabilities existing between both parties on the expiry of the validity of the contract shall not be influenced by the expiration of this contract. The debtor shall be kept liable until the debtor fully pays up his debts to the creditor.

    本合同期限届满时，双方发生的未了债权和债务不受合同期满的影响，债务人应向债权人继续偿付未了债务。

5. Both parties shall abide by/All the activities of both parties shall comply with the contractual stipulations.

    双方或双方的一切活动都应遵守合同规定。

6. This Contract is made by and between the Buyer and the Seller, whereby the Buyer agrees to buy and the Seller agrees to sell the under-mentioned commodity subject to the terms and conditions stipulated below.

买卖双方同意按下述条款购买/出售下列商品并签订本合同。

7. At the request of Party B, Party A agrees to send technicians to assist Party B to install the equipment.

   应乙方要求,甲方同意派遣技术人员帮助乙方安装设备。

8. The Employer shall render correct technical guidance to the personnel.

   雇主应该对有关人员给予正确的技术指导。

9. Party A shall repatriate the patient to China and bear the cost of his passage to Guangzhou.

   甲方应将病人遣返中国并负责其返回广州的旅费。

10. This Contract shall be governed by and construed in accordance with the laws of China.

    本合同受中国法律管辖,并按中国法律解释。

11. The Employer may object to and require the Contractor to replace forthwith any of its authorized representatives who is incompetent.

    雇主认为承包人委派的授权代表不合格时,可以反对并要求立即撤换。

12. The Chairperson may convene an interim meeting based on a proposal made by one-third of the total number of directors.

    董事长可以根据董事会过1/3董事的提议而召集临时董事会议。

13. In case one party desires to sell or assign all or part of its investment subscribed, the other party shall have the preemptive right.

    如一方想出售或转让其投资之全部或部分,另一方就有优先购买权。

14. In processing transactions, the manufacturers shall never have title either to the materials or the finished products.

    加工贸易中,厂方无论是对原料还是成品都无所有权。

15. The term "Effective date" means the date on which this Agreement is duly executed by the parties hereto.

    "生效期"指双方合同签字的日子。

# UNIT 6

## TOPICS

- ➢ What is a computer network?
- ➢ When did the first computer network arise?
- ➢ What is the Internet?
- ➢ What is the World Wide Web?
- ➢ What did the Internet bring to us?
- ➢ Bargaining skills.
- ➢ How to write a Lost and Found Notice?

# Computer Network and Internet

**Simply put**, a computer network is a collection of **autonomous** computers. If you think about computer network, then you should consider the fact that it simple means various computers connected to each other which should facilitate sharing of resources.

Computer networks are often classified as local area network (LAN), wide area network (WAN), **metropolitan** area network (MAN), Wireless Networks and Internetworks.

The Internet is a giant network of computers located all over the world that communicate with each other.

The Internet is an international collection of computer networks that all understand a standard system of addresses and commands, connected together through **backbone** systems. It was started in 1969, when the **U.S. Department of Defense** established a nationwide network to connect a handful of universities and contractors. The original idea was to increase computing capacity that could be shared by users in many locations and to find out what it would take for computer networks to survive a **nuclear** war or other disaster by providing multiple paths between users. People on the **ARPANET** (as this nationwide network was originally called) quickly discovered that they could exchange messages and conduct electronic "conferences" with distant colleagues for purposes that had nothing to do with the **military industrial complex**. If somebody else had something interesting stored on their computer, it was a simple matter to obtain a copy (assuming the owner did not protect it).

Over the years, additional networks joined which added access to more and more computers. The first international

simply put 简言之
autonomous [ɔː'tɔnəməs] adj. 自治的；自主的；自发的

metropolitan [ˌmetrə'pɔlitən] adj. 大都市的；大主教辖区的；宗主国的 n. 大城市人；大主教；宗主国的公民

backbone ['bækbəun] n. 支柱；主干网；决心，毅力；脊椎
U.S. Department of Defense 美国国防部

nuclear ['njuːkliə] adj. 原子能的；中心的；细胞核的；原子核的
ARPA (Advanced Research Projects Agency) 美国国防部高级研究计划局
military industrial complex 军工联合企业 军事工业复合体

connections, to Norway and England, were added in 1973. Today thousands of networks and millions of computers are connected to the Internet. It is growing so quickly that nobody can say exactly how many users "On the Net".

The Internet is the largest **repository** of information which can provide very large network resources. The network resources can be divided into network facilities resources and network information resources. The network facilities resources provide us the ability of remote computation and communication. The network information resources provide us all kinds of information services, such as science, education, business, history, law, art, and entertainment, etc.

repository [ri'pɔzitəri] n. 贮藏室，仓库；知识库；智囊团

The goal of your use of the Internet is exchanging messages or obtaining information. What you need to know is that you can exchange message with other computers on the Internet and use your computer as a **remote terminal** on distant computers. But the internal details of the link are less important, as long as it works. If you connect computers together on a network, each computer must have a unique address, which could be either a word or a number. For example, the address of Sam's computer could be Sam, or a number.

remote [ri'məut] n. 远程 adj. 遥远的；偏僻的；疏远的
terminal ['tə:minəl] n. 末端；终点；终端机；极限 adj. 末端的；终点的；晚期的

The Internet is a huge interconnected system, but it uses just a handful of method to move data around. Until the recent **explosion** of public interest in the Internet, the **vast majority** of the computers on the Net use the Unix operating system. As a result, the standard Unix commands for certain Internet services have entered the online community's languages as both nouns and verbs to describe the services themselves. Some of the services that the Internet can provide are: Mail, Remote use of another computer (Telnet), File Transfer Protocol (FTP), News, and Live conversation.

explosion [ik'spləuʒən] n. 爆炸；爆发；激增
vast majority 绝大多数，大部分

The most commonly used network service is electronic mail (E-mail), or simply as mail. Mail permits network users to send textual messages to each other. Computers and networks handle delivering the mail, so that communicating mail users do not have to handle details of delivery, and do not have to be present at the same time or place.

Presently, a user with an account on any Internet machine can establish a live connection to any other machine on the Net from the terminal in his office or laboratory. It is only necessary to use the Unix command that sets up a remote terminal connection (Telnet), followed by the address of the distant machine. Before you can use the Internet, you must choose a way to move data between the Internet and your PC. This link may be a high-speed data communication circuit, a local area network (LAN), a telephone line or a radio channel. Most likely, you will use a **Modem** attached to your telephone line to talk to the Internet. Naturally, the quality of your Internet connection and service, like many other things in life, is dictated by the amount of money that you are willing to spend.

modem ['məudem] n. 调制解调器

The simplest way to access a file on another host is to copy it across the network to your local host. FTP can do this.

Although all these services can well satisfy the needs of the users for information exchange, a definite requirement is needed for the users. Not only should the users know where the resources locates, but also he should know some operating commands concerned to ease the searching **burden** of the users, recently some convenient searching tools appears, such as WWW.

burden ['bə:dən] n. 负担；责任；船的载货量 vt. 使负担；烦扰；装货于

WWW (World Wide Web) is a networked **hypertext** protocol and user interface. It provides access to multiple services and documents like **Gopher** does, but is more **ambitious** in its method. A jump to other Internet service can be triggered by a mouse click on a "hot link" word, image, or icon on the Web page.

hypertext ['haipətekst] n. [计] 超文本(含有指向其他文本文件链接的文本)

Gopher['gəufə] n. 一种由菜单式驱动的信息查询工具

ambitious [æm'biʃəs] adj. 野心勃勃的；有雄心的；热望的；炫耀的

WWW is the most popular part of the Internet by far. Once you spend time on the Web, you will begin to feel like there is no limit to what you can discover. The Web allows rich and **diverse** communication by displaying text, graphics, animation, photos, sound and video.

diverse [dai'və:s] adj. 不同的；多种多样的；变化多的

So just what is this **miraculous** creation? The Web physically consists of your personal computer, web browser software, a connection to an Internet service provider, computers called servers that host digital data and routers and switches to direct the flow of information.

miraculous [mi'rækjuləs] adj.
不可思议的，奇迹的

As more and more systems join the Internet, and as more and more forms of information can be converted to digital form, the amount of stuff available to Internet users continues to grow. At some points very soon after the nationwide (and later worldwide) Internet started to grow, people began to treat the Net as a community, with its own tradition and customs. For example, somebody would ask a question in a conference, and a complete stranger would send back an answer. After the same question were repeated several time by people who hadn't seen the original answers, somebody else gathered list of "frequently asked questions" and placed it where newcomers could find it.

So we can say that the Internet is your PC's window to the rest of the world.

## EXERCISES

Ⅰ. Match the terms and the interpretations.

1. Backbone  (a) A multitasking, multi-user computer operating system originally developed in 1969 by a group of AT&T employees at Bell Labs.

2. ARPANET  (b) A network protocol used on the Internet or local area networks to provide a bidirectional interactive text-oriented communications facility using a virtual terminal connection.

3. Telnet  (c) The part of a communication network that carries the heaviest traffic.

4. Unix  (d) It was the world's first operational packet switching network and the core network of a set that came to compose the global Internet.

5. Remote Terminal  (e) A terminal connected to a computer by a data link.

II. Are the following statements True (T) or False (F)?

1. (    ) A computer network is a collection of autonomous computers.

2. (    ) Development of the network began in 1950s.

3. (    ) The Internet is a giant network of computers located all over the world that communicate with each other.

4. (    ) The first network in the world is Internet.

5. (    ) The first international connections, to Norway and England, were added in 1973.

6. (    ) FTP is the most popular part of the Internet by far.

7. (    ) The most commonly used network service is electronic Telnet.

8. (    ) FTP can copy a file on another host to your local host across the network.

9. (    ) Computers and networks handle delivering the mail, so that communicating mail users do not have to handle details of delivery.

10. (    ) Live conversation is a service that the Internet can not provide.

III. Translate the following words and phrases into Chinese.

1. Autonomous Computers          _____

2. LAN                            _____

3. Wireless Network               _____

4. Network Facilities Resources   _____

5. Interconnected System          _____

6. FTP                            _____

7. Modem                          _____

8. WWW                            _____

9. Digital Form                   _____

10. Nationwide                    _____

IV. Translate the following Chinese statements into English.

1. 万维网是一个基于超文本的系统,可用于查询和访问互联网资源。

2. 我认为您的调制解调器现在已经无法使用了。
_____

3. 在当今，互联网的应用无处不在，我们的工作和生活都离不开它。
_____

4. 电子邮件是一种便捷又经济的发送信息的方式。
_____

5. 互联网上的信息都是以多媒体文件的形式显示出来的。
_____

Ⅴ. Fill in each of the blanks with one of the following words or phrases.

*equip   overall   service   data sources   router*

*associate with   connect   wireless   corporation   process*

Computer networking devices are any types of devices that are_____a computer network and aid in the_____of managing data. Also known as network equipment or interworking units, the computer networking device may be some type of hardware equipment or a _____ that aids in the processing of data in some capacity. Any type of network, ranging from a simple home network to a wide area network (WAN) utilized by a large_____will make use of various networking devices.

One of the more basic examples of a computer networking device is the_____. This type of device makes it possible to connect with_____and share information over the applications used within the_____network structure. One of the more common examples of a router is a dual router used in many home networks that is_____directly to the master computer that serves as the server for that network, and also is_____to provide a _____connection to other computers in the home.

# SUPPLEMENTARY

## Commonly Used Computer Network

You have probably heard of networks and networking, and you may even be using one right now, what exactly is a computer network? And why would someone want to set one up or be part of one?

### What is a Computer Network?

A network is basically a set of two or more articles that are linked so the computers can share resources, such as printers, software, and internet connections. Networked computers can also share files without having to transfer data using a disk or data key. And users of networked computers can also communicate electronically without use of the internet.

Computers within a network can be linked several ways: cables, telephone lines, radio waves, satellites, or infrared beams. There are also three basic types of networks: Local Area Network (LAN), Metropolitan Area Network(MAN), and Wide Area Network(WAN).

## Local Area Networks

A Local Area Network (LAN) is basically a smaller network that's confined to a relatively small geographic area. LAN computers are rarely more than a mile apart. Examples of common LANs are networked computers within a writing lab, school, or building.

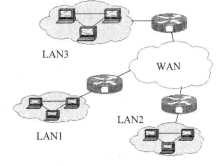

Within a LAN network, one computer is the file server. This means that it stores all software that controls the network, and it also stores the software that can be shared among computers in the network. The file server is the heart of the LAN.

The computers attached to the file server are called workstations. Workstations can be less powerful than the file server because they don't have to store as many files and applications as the file server, and they are not always on and working to keep the network up and running. However, workstations may also have additional software stored on their hard drives. Most LANs are connected using cables.

## Metropolitan Area Networks

A Metropolitan Area Network (MAN) connect 2 or more LANs together but does not span outside the boundaries of a city, town, or metropolitan area. Within this type of network is also the Campus Area Network (CAN), which is generally smaller than a MAN, connecting LANs within a limited functional area, like a college campus, military base, or industrial complex.

## Wide Area Networks

Wide Area Networks connect larger geographic areas. Often, smaller LANs are interconnected to form a large WAN. For instance, an office LAN in Los Angeles may be connected to office LANs for the same company in New York, Toronto, Paris, and London to form a WAN spanning the whole company. The individual offices are no longer part of individual LANs, they are instead part of a worldwide WAN.

The connection of this type of network is complicated. WANs are normally connected using multiplexers connect local and metropolitan networks to global communications networks like the Internet.

# CONVERSATION

## Bargaining Skills

## 议价的技巧

议价即讨价还价，在日常生活中，讨价还价是人人都会遇到的问题。怎样议价效果才最好呢？当对方报价后，如何说服对方提供更优惠的价格呢？议价时需注意以下几点原则：

1. 漫不经心，声东击西

当你看好某商品时，不要急着问价，先随便问一下其他商品的价格，表现出很随意的样子，然后突然问你要的东西的价格。店主通常不及防范，报出较低的价格。切忌表露出对你想要的那件商品的热情，善于察颜观色的店主会漫天起价。

2. 以理服人，见好就收

因讨价还价是伴随着价格评议进行的，故应尊重对方。议价应有理有据，切勿不给出充分理由而强压对方报价，使议价陷入僵局。如对方同意降价，达到自己心里价位或接近心理价位即可，不易强求，应保持平和的心态。

3. 揣摩心理，掌握次数

如遇到漫天要价者，应不动声色，揣摩对方心理。还价时，先压价至自己的心理价位并给出原因，再观察对方反应，逐步提价。提价时应注意尺度，可观察对方态度，随机调整提价额度。

4. 评头论足，欲擒故纵

试着用最快的速度列举出该货品的缺点，一般的顺序是式样、颜色、质地、做工，从而达到减价的目的。对方若不肯减价，这时即使你真的喜欢这件商品，也要当做不在意，转身就走，迫使对方减价。

## Example

### Dialogue 1

A: Hi, how much do you want for this?
B: 150 yuan.
A: What? Don't you try to rip me off! I know what this is cost, 40 yuan, tops.
B: No way! It cost me more than that, 120!
A: Come on! If you don't give a better price, I won't buy this from you.

| | |
|---|---|
| B: | 110. Take it or leave it. |
| A: | I'll give you 55. |
| B: | I can't do that. I have to make a living. Give me 100 and it's yours. |
| A: | That's still much too expensive. (Starting to walk away) |
| B: | Wait, wait! Ok, 85, final price. |
| A: | If that's the lowest you're willing to go. I'm leaving. I'll pay 65, final offer. |
| B: | You drive a hard bargain. I'm losing money on this, but alright. I'll let you have it for 65. |
| A: | Thanks a lot. |

## ▷ Dialogue 2

| | |
|---|---|
| A: | Good afternoon. Can I help you? |
| B: | I need a dinette set. |
| A: | How's this one? |
| B: | Looks all right. How much do you sell it? |
| A: | 850 dollars. |
| B: | This is out of my price range. |
| A: | What's your general price range? |
| B: | Around 300 dollars. |
| A: | Is this one all right? It costs 380 dollars. |
| B: | It suits me. Does the price include delivery and installation charge? |
| A: | Yes, of course. |
| B: | Is the price negotiable? |
| A: | I am so sorry. No bargaining. |
| B: | Come on, give me a discount. If you don't give me a better price, I won't buy this. |
| A: | All right, I'll give it to you for 350 dollars. |
| B: | OK .Thank you so much. I'll take this one. |

## Practice

Imagine that you want to buy a T-shirt and your partner is the seller. You hope the seller give you a discount. Practice bargaining skills with the given points:

- ◇ Quote the price.
- ◇ Dissent from the price with your reasons.
- ◇ Ask for a discount.
- ◇ Tell the seller your bottom line.
- ◇ Make an agreement.

# Tips

## ☒ 讨价还价常用语

1. Don't try to rip me off. I know what this cost. 别想宰我，我识货。
2. Can you give me a little deal on this? 这件东西能卖得便宜一点吗？
3. Can you give me this for cheaper? 能便宜一点给我吗？
4. Is there any discount on bulk purchases? 我多买些能打折吗？
5. Give me a reduction in price, please. 给我打个折吧。
6. How much do you want for this? 这件东西你想卖多少钱？
7. If you don't give me a better price, I won't buy this. 如果价格不更优惠些，我是不会买的。
8. I can get this cheaper at other places. 这件东西我在别的地方可以买到更便宜的。
9. What's the lowest you're willing to go? 最低你能出什么价？
10. Come on, give me a break on this. 别这样，你就让点儿价吧。
11. Could you give me a discount? 能给我个折扣吗？
12. Are these clothes on sale? 这些衣服打特价吗？
13. Is the price negotiable? 这个价钱可以商量吗？
14. How about buy one and get one free? 买一送一怎么样？
15. Can you give me a better deal? 可以给我更好的价钱吗？
16. I'd buy it right away if it were cheaper. 便宜一点的话我马上买。
17. The price is beyond my budget. 这个价钱超出我的预算了。
18. I'll give 500 dollars for it. 五百块我就买。
19. That's steep, isn't it? 这个价钱太离谱了吧？
20. It's too expensive. I can't afford it. 太贵了，我买不起。

WRITING

## How to Write a Lost and Found Notice
## 怎样写寻物启事及失物招领

在英语中，寻物启事及失物招领都可以直接译成 Lost and Found，并没有严格的区分。按照字面上的意思，寻物启事可对应 Lost，失物招领可对应 Found。

1. 寻物启事(Lost)

寻物启事一般包括以下几项内容。
(1) 标题。寻物启事的标题可以有两种构成格式:
第一,由文种名称和缘故构成。如"寻物启事"。
第二,由文种名和具体丢失物名构成。如"寻书启事"、"寻自行车启事"。
(2) 正文。寻物启事的正文一般由以下几项内容构成:
其一,写明丢失物的名称、外观、规格、数量、品牌等,同时要写明丢失的原因、时间和具体地点。
其二,交代清楚拾物者送还的具体方式,或注明发文者的详细地址、联络方式等。
第三,寻物启事是求人协助寻找的,故除文中写些表谢意的话外,还可以写明给以拾到者必要的酬金之类的话。
(3) 落款。落款要署上发文的单位或个人的名称或姓名,并署上发文的日期。
2. 失物招领(Found)

失物招领并没有硬性的格式规范,而且内容一般都很简练。但需要注意的是,对于拣到的东西说明不能太详细,以防止有人冒领,同时写清楚联系人的联系方式,确保失主可以联系到您。

# Example

**LOST**

Mr. White carelessly lost his suitcase at 9:30 a.m. this Saturday when he took a bus from Datong to Taiyuan.

It is an orange square leather suitcase with a metal handle on it. A label with White's name is tied to the handle. Inside the suitcase are two jackets and a camera. There is a Chinese-English dictionary and a letter from America in the packet on the front cover and in the back packet is a wallet with 1,000 dollars and a train ticket from Taiyuan to Beijing inside.

Will the finder get in touch with Mr. White, please? His telephone number is 13931811751. Mr. White will appreciate the finder very much.

Taiyuan Bus Station

10th September, 2009

## 寻 物 启 事

　　本周六上午九点半，怀特先生乘汽车从大同来太原时，不慎将手提箱丢失。

　　手提箱为橘色、方形、皮制，其上有一个金属提手。左提手上系着一个写有怀特名字的标签。箱内有两件上衣和一部相机。箱前的口袋里有一本汉英词典和一封来自美国的信。箱后口袋里有一个钱包，里面装有1000美元和一张从太原到北京的火车票。

　　请拾到者与怀特先生联系，他的电话号码是13931811751。怀特先生非常感激拾到者。

太原汽车站

2009年9月10日

---

**FOUND**

A bunch of keys has been found in our school canteen today after lunch time. The owner is advised to contact Jane at 18655261853 to claim the keys within a week.

Lost and Found Office of Computer Department

20th September, 2011

---

## 失 物 招 领

　　Jane 于今天中午午餐后在学校餐厅捡到钥匙一串，请失主于一周内与 Jane 联系并认领失物，电话是18655261853。

计算机系失物招领处

2011年9月20日

## Practice

**LOST**

_____
_____
_____

Loser,
Jack Stone
15th May, 2010

✉ <center>**寻 物 启 事**</center>

　　今丢失红色公文包一只，公文包上有包含失主信息的名牌，失主的电子邮箱也写在名牌上了，公文包有些损坏。请拾到者通过邮件或电话联系失主，电话是010-80001234，失主愿意支付50美元以示酬谢。

失主：杰克·斯通<br>2010年5月15日

---

**FOUND**

_____

_____

_____

_____

Finder,
Mary Brown
7th June, 2011

---

✉ <center>**失 物 招 领**</center>

　　昨天下午在操场拾到书包一个，内有几本图书、一个铅笔盒和一块手表，请失主直接联系玛丽·布朗认领失物，联系电话021-84729646。

拾到者：玛丽·布朗<br>2011年6月7日

## Tips

✉ 当我们发现丢失了物品时，一般会用什么样的方式来表达呢？

A:　Excuse me. I've got a problem. I think I've lost my wallet.

B:　Are you sure it's not in your bag?

A: Yes. I've looked for it.

B: Why don't you check the Lost Property Office?

A: That's a good idea. Thank you.

> 当我们发现物品丢失时，大多会想到去失物招领处寻找。如果经过询问后还是未找到，我们将怎么做呢？

Kate: Excuse me, could you help me?

Clerk: Yes. What seems to be the problem?

Kate: Well, I was wondering if anyone has turned in a passport.

Clerk: I'm afraid not. Have you lost your passport?

Kate: I think so. I can't find it anywhere in my hotel room, and I remember the last place I used it yesterday was in this department store.

Clerk: Where exactly did you use your passport in the store?

Kate: In the suit-dress department. I had to show it to pay for these dresses with my traveler's checks.

Clerk: Well, let me call the suit-dress department to see if they've found a passport.

(A minute later)

Clerk: Sorry. Your passport is not been turned in there, either.

Kate: Then what shall I do?

Clerk: You can fill in this lost property report, and I'll keep my eye out for it. Those kinds of things usually turn up eventually, but I suggest you contact your embassy and tell them about your situation, so they can issue you a new passport in case it doesn't show up.

Kate: You're right. Do you have a pen?

Clerk: Here you are.

Kate: Oh, I seem to lose something every time I travel.

> 当失主看到失物招领启事或听到失物招领的广播后，应立即联系认领，在认领的过程中一般都会被问及失物的细节，失主对此应有所准备。

Clerk: Can I help you, sir?

Benjamin: I'm here for the backpack you announced several minutes ago.

Clerk: OK, take a seat please, sir. First of all, can you show me your ID card please?

Benjamin: Sure. Here you are.

Clerk: OK, could you please tell me what your backpack looks like?

Benjamin: Of course, it's a soft leather one, you know, not a sports one that looks childish.

Clerk: Mmm, does it zip closed?

Benjamin: No, its straps closed, and it has a buckle in the front.

Clerk: OK, can you tell me the distinguishing features of this backpack?

Benjamin: Oh, yeah, the brand name.

Clerk: So what's it, sir?

Benjamin: Oh, it's Polo. It has the logo on the back and at the bottom in the left-hand corner.

Clerk: OK, can you name the items in it?

Benjamin: Well, all the gifts for my family, you know two pairs of sneakers for my children and a bottle of perfume for my wife.

Clerk: OK, sir, I'm sure it's your bag. Thank you for your cooperation. You can have it now.

Benjamin: Thank you so much. You guys are really responsible.

# UNIT 7

# TOPICS

- What is the definition of ecommerce?
- What is B2B?
- What is the difference between B2B and B2C?
- Do you know something about Amazon?
- What is online shopping?
- What methods of payment could be used in ecommerce?
- What do you think of the future of ecommerce?
- How to place an order?
- How to write letter of complaint and letter of apology?

TEXT

# E-commerce (electronic commerce or EC)

Electronic commerce, commonly known as e-commerce, **ecommerce**, or e-comm, **consists of** the buying and selling of products or services over electronic systems such as the Internet and other computer networks, to generate **revenue** or support revenue generation. In practice, this term and a newer term, e-business, are often used **interchangeably**. For online retail selling, the term e-tailing is sometimes used.

Ecommerce covers **a range of** different types of businesses, from consumer based retail sites, through **auctions** or [legal, paid for] downloading of music and movies, to business exchanges trading goods and services between corporation. It is currently become one of the most important aspects of the internet to **emerge**.

Electronic commerce that takes place between businesses is **referred to** as business-to-business or B2B rather than between a business and a consumer. B2B businesses often deal with hundreds or even thousands of other business, either as customers or suppliers. Carrying out these transactions electronically provides vast competitive advantages over traditional methods. It is faster, cheaper, and more convenient and can reach more **potential** customers. Electronic commerce that takes place between businesses and consumers, on the other hand, is referred to as business-to-consumer or B2C. This is the type of electronic commerce conducted by companies such as **Amazon**.com, a retailer based in Seattle, Washington. This company does not have physical stores, but sells all their books and other products over the intent. What was called **Online shopping** is a form of electronic commerce

| | |
|---|---|
| ecommerce | 电子商务 |
| consist of | 由……组成 |
| revenue ['revinju:] n. | 税收, 收入, 税务局 |
| interchangeably [ˌintə'tʃeindʒəbli] adv. | 可交换地 |
| a range of | 一套, 一系列 |
| auction ['ɔ:kʃən] n. | 拍卖 |
| emerge [i'mə:dʒ] vi. | 浮现, (由某种状态)脱出, (事实)显现出来 |
| refer to | 涉及, 指的是, 提及, 参考, 适用于 |
| potential [pə'tenʃəl] adj. | 潜在的, 可能的 n. 潜力, 潜能; 电位, 电势 |
| Amazon ['æməzən] n. | 亚马逊(全球网商零售巨擘) |
| Online shopping | 网上购物 |

where the buyer is directly online to the seller's computer usually **via** the internet with no barriers of time or distance.

The amount of trade conducted electronically has grown **extraordinarily** with widespread Internet usage. The use of commerce is conducted in this way, **spurring** and drawing on **innovations** in electronic funds transfer, supply chain management, Internet marketing, online transaction processing, electronic data interchange (EDI), inventory management systems, and **automated** data collection systems. Originally, electronic commerce was identified as the **facilitation** of commercial transactions electronically, using technology such as Electronic Data Interchange (EDI) and Electronic Funds Transfer (EFT). As the growth and acceptance of credit cards, automated teller machines (ATM) and telephone banking were also became forms of electronic commerce. Later, a worldwide system, which called internet/www, transformed an **academic** telecommunication network into everyman everyday **communication.** Modern electronic commerce typically uses the World Wide Web at least at one point in the transaction's life-cycle, although it may encompass a wider range of technologies such as e-mail, mobile devices and telephones as well. By the end of 2000, people began to **associate** a word "ecommerce" with the ability of purchasing various goods through the Internet using secure **protocols** and electronic payment services.

As a retailer, to create a successful online store can be difficult if unaware of ecommerce **principle**. Researching and understanding the guidelines required to properly **implement** an ecommerce plan is a crucial part to becoming successful with online store building. With the help of good marketing strategy, it will driving targeted traffic to your site and a means of **enticing** returned customer. As a customer, you need a means of accepting payments. This usually entails obtaining a merchant account and accepting credit cards through an online Pauline gateway. We can also pay through our bank card,

via ['vaiə] prep. 经由，通过

extraordinarily [iks'trɔ:dnrili] adv. 非常(格外)
spur [spə:] vt. 刺激，鞭策，促进
innovation [,inəu'veiʃən] n. 创新，革新

automate ['ɔ:təmeit] v. 使自动化

facilitation [fə,sili'teiʃn] n. 简易化，促进，使人方便的东西

academic [,ækə'demik] adj. 学院的，学术的，理论的
communication [kə,mju:ni'keiʃən] n. 交流，通讯，传达，通信，沟通

associate [ə'səuʃieit] vt. 联想，联合

protocol ['prəutəkɔl] n. 外交礼仪，草案，协议，规章制度

principle ['prinsəpl] n. 原理，原则，主义，信念
implement ['implimənt] n. 工具，器具，当工具的物品

entice [in'tais] v. 诱使，引诱

**Alipay, Tenpay**, etc. **Cash on arrival** is acceptable too. Nowadays, more and more people choose online shopping as the most convenient way to get what they want. In some special days, it will give a lot discounts and **price reduction.**

Alipay   n. 支付宝
Tenpay   n. 财付通
cash on arrival   货到付款
price reduction   n. 削价, 降价

Economists have **theorized** that ecommerce ought to lead to intensified price competition, as it increases consumers' ability to gather information about products and prices. It was found that the growth of online shopping has also **affected** industry structure in two areas that have seen **significant** growth in ecommerce, bookshops and travel agencies. Generally, larger firms have grown at the expense of smaller ones, as they are able to use economies of scale and offer lower prices.

theorize ['θiəraiz] v. 推理, 建立理论, 理论化=theorise

affect [ə'fekt] vt. 影响, 作用
significant [sig'nifikənt] adj. 重要的, 有意义的, 意味深长的

Business models across the world also continue to change **drastically** with the advent of ecommerce. Amongst emerging economies, China's ecommerce presence continues to **expand**. With 384 million internet users, China's online shopping sales rose to $36.6 billion in 2009 and one of the reasons behind the huge growth has been the improved trust level for shoppers. The Chinese retailers have been able to help consumers feel more comfortable shopping online. E-commerce has become an important tool for business worldwide not only to sell to customers engage them. It is expected to **boom** beyond limits in near future, and it will **play a major role** in the way that all the companies, large or small, conduct business either with their customers, other businesses, or both.

drastically ['dræstikəli] adv. 大幅度地, 彻底地
expand [iks'pænd] v. 使……膨胀; 详述; 扩张, 增加, 张开

boom [bu:m] vi. 急速增长 vt. 使兴旺, 促进 n. 繁荣, 兴旺
play a major role   起重要作用

 EXERCISES

Ⅰ. Match the terms and the interpretations.

1. Amazon.com      (a) A way for companies and banks to send information to each other by computer using an agreed FORMAT so that the company receiving the documents can easily read them on their computer and print them out on paper.

2. EDI  (b) A machine outside a bank that you use to get money from your account.

3. ATM  (c) A US website that sells books, music, toys etc. You make your order and pay over the Internet, and the books, toys etc are sent to you through the post. It also has websites in other countries such as the UK and Germany.

4. Protocol  (d) To make it easier for a process or activity to happen.

5. Facilitate  (e) An established method for connecting computers so that they can exchange information.

II. Are the following statements True (T) or False (F)?

1. (   ) Ecommerce refers to B2B.

2. (   ) Nowadays, credit card is the only acceptable way of ecommerce.

3. (   ) More and more people choose purchase online because it is the most convenient way.

4. (   ) To start a successful online store can be difficult if you have no money.

5. (   ) Smaller firms can able use economies of scale and offer lower prices.

6. (   ) Ecommerce has become an important tool just for business.

7. (   ) Compared with traditional stores, online shopping can earn more money.

8. (   ) By the end of 2000, people began to use internet to purchase what they want online.

9. (   ) Economists thought that consumers can lead to intensified price competition.

10. (   ) In the near future, ecommerce will play a major role just in larger company.

III. Translate the following words and phrases into Chinese.

1. Mobile Commerce  _____

2. Collaborative Commerce  _____

3. Online Shopping  _____

4. Electronic Data Interchange  _____

5. Enterprise Resource Planning  _____

6. Material Request Plan  _____

7. Cash on Arrival  _____

8. B2C(Business To Customer)  _____

9. E-Business Platform　　　　_____

10. Third Party Logistics　　　　_____

Ⅳ. Translate the following Chinese statements into English.

1. 这些发生在企业与企业之间的电子商务活动被称为是企业对企业的，即 B2B，而不是指企业对消费者。

   _____

2. B2B 也比这种传统的方式更快、更便宜、更方便，而且可以争取到更多潜在客户。

   _____

3. 这个公司不仅拥有实体店，而且通过网络销售他们的书籍和其他产品。这种方式就被称为网上购物。

   _____

4. 作为一个零售商，如果没有意识到电子商务的规则而去成功创建一个网上商店也许会有点困难。

   _____

5. 电子商务的发展不可估量。各类公司，不论其规模大小，在同客户及其他企业进行交易，或与客户和企业同时进行交易的时候，电子商务必将发挥举足轻重的作用。

   _____

Ⅴ. Fill in each of the blanks with one of the following words or phrases.

*consist of　　associate　　innovation　　automate　　spur　　boom　　affect*

*expand　　principle　　refer to*

1. Metals _____ when they are heated.

2. International competition was a _____ to modernization.

3. It cans_____ an academic subject or a practical skill.

4. The United Kingdom _____ Great Britain and Northern Ireland.

5. Your behavior does not accord with your_____.

6. I'm very selective about the people I _____ with.

7. Don't let this trifling matter _____ our harmonious relations.

8. He made his pile during the property_____.

9. Every one of us is active in technical_____.

10. Many banks have begun to _____.

# SUPPLEMENTARY

## About Taobao

Taobao is a Chinese language web site for online shopping, similar to eBay, Rakuten(日本乐天) and Amazon, operated in the People's Republic of China by Alibaba Group.

Founded by Alibaba Group in May 10th 2003, Taobao facilitates business-to-consumer (B2C) and consumer-to-consumer (C2C) retail by providing a platform for businesses and individual entrepreneurs to open online retail stores that mainly cater to consumers in mainland China, Hong Kong, Macau and Taiwan.

### History

Taobao was launched in May 2003 by Alibaba Group after eBay acquired Eachnet, China's online auction leader at the time, for US $180 million and became the major player in the Chinese consumer e-commerce market. To counter eBay's expansion,  Taobao offered free listings to sellers and introduced website features designed to better cater to local consumers, such as an instant messaging tool for facilitating buyer-seller communication and an escrow-based payment tool, Alipay. As a result, Taobao became the undisputed market leader in mainland China within two years. Its market share jumped from 8% to 59% between 2003 and 2005, while eBay China's slid from 79% to 36%. eBay had to shut down its own site in China in 2006.

In April 2008, Taobao introduced a dedicated B2C platform called Taobao Mall to complement its C2C marketplace. Since then, Taobao Mall has established itself as the destination for quality, brand name goods for Chinese consumers. Taobao Mall launched an independent web domain, tmall.com, in November 2010, and changed its Chinese name to Tian Mao (Tmall) on January 2012. The site currently features products from more than 30,000 major multinational and Chinese brands.

In October 2010, Taobao beta launched eTao as an independent online shopping search engine, providing product and merchant information from all major consumer e-commerce websites in China.

In June 2011, Alibaba Group Chairman and CEO Jack Ma revealed that Taobao will be split into three different companies: eTao (to be used for shopping searches), Tmall (a B2C platform), and Taobao Marketplace(a C2C platform).

## Taobao Marketplace

Taobao Marketplace is the platform for Consumer-to-consumer e-commerce, the main part of Taobao. Sellers are able to post commodities on the Taobao Marketplace either through a fixed price or by auction. The overwhelming majority of the products on Taobao are brand new merchandise sold at a fixed price; auctions make up a very small percentage of transactions. Buyers can judge the sellers credit from their selling prestige, or the history of comment and complaint.

## Tmall

Tmall was established as Taobao Mall within Taobao in April 2008 as a dedicated B2C platform within Taobao. In November 2010, it introduced an independent web domain, tmall.com, to  differentiate listings by Taobao Mall merchants, who are either brand owners or authorized distributors, from Taobao's C2C merchants. On January 11, 2012, Taobao Mall officially changed its Chinese name to Tian Mao (天猫), the Chinese pronunciation of Tmall.

Brands that have established flagship stores on Tmall include P&G, adidas, UNIQLO, GAP, Nine West, Reebok, Ray-Ban, New Balance, Umbro, Lenovo, Dell, Nokia, Philips, Samsung, Logitech and Lipton.Alipay.

## AliWangWang

 A distinctive feature of shopping on Taobao is the pervasive communication between buyer and seller prior to the purchase through its embedded proprietary instant chat program, named AliWangWang. It has become a habit among Chinese online shoppers to "chat" with the sellers or their customer service team through AliWangWang to inquire about products, engage in bargaining prior to purchase products.

## Happy Taobao

In Dec. 2009, Taobao together with Hunan TV set up Happy Taobao, Inc for television shopping. Not only did Hunan TV launch an entertainment called "Happy Taobao", Taobao Marketplace created channel and independent website, both of which were combine electronic commercial with TV media.

**Financials**

In 2009 Taobao saw advertising revenues of $220 million. Advertising accounts for a majority of Taobao's income. Analysts believe Taobao's turnover exceeded 400 billion RMB in 2010 and revenues surpassed 5 billion RMB.

In January 2010, Alibaba Group said it expects gross merchandise volume on Taobao to double from 200 billion Chinese yuan in 2009 to 400 billion Chinese yuan in 2010, as China's e-commerce market is expected to grow significantly in the next five to eight years.

## How to Place an Order

# 如何下订单

订货(order)是买方为要求供应具体数量的货物而提出的一种要求。此时，交易双方之间的陌生感已消除，可以说已经度过了接触障碍和难关。下订单时应注意以下几点：

- ➢ 开始交谈时就直接说明订购的意图。

- ➢ 下订单时应交代清楚商品的名称、品质、数量、包装、价格条件、支付条件以及需要对方提供的单据等各项信息。内容务必准确、清楚，否则会带来不必要的损失与麻烦。

- ➢ 若卖方无法提供买方所需要的货物，则最好介绍一些合适的替代品；若买方所需货物的价格和规格发生了变化，卖方可提出还价并劝买方接受。尤其要注意：当卖方拒绝接受订单时，必须非常谨慎，应为日后有可能的其他交易留下余地。

# Example

## ⊠ Dialogue 1

A: Our toner cartridges are already out of ink, could you make an order for a new set?

B: We will need new cartridges for all of the office printers? That will be a large order, probably about two or three cases. The office supply store we usually go through might not have that many in stock.

A: You can double check with the housekeeping department, but I am pretty sure all of the machines will need new cartridges. Last time when we made our order to the supplier, the quantity was also especially high. They are used to receiving such bulk orders from us. As long as we give them a heads up a couple days in advance, they can usually fill the order.

B: OK, I will make a few calls and run our order by housekeeping first to make sure. Is there anything else we need to order while I am at it?

A: I think the only thing is toner. Try to see if they can deliver it before the end of business day tomorrow. We should really try to do better about waiting until the last minute to fill orders that are usually made on a monthly basis. Anyhow, see what you can do to expedite the order this time.

B: OK, will do.

## ⊠ Dialogue 2

A: May I help you?

B: Yes, I would like to place an order for toner cartridges. We have a standing agreement with your company, so we will need the same amount as last time.

A: Let me key in your information into my computer. I will pull up our records for you. Do you have an order number? What name is the order listed under?

B: It should be under Leslie Smith. The number is 184796 A.

A: Yes, Mr. Smith. I have an order for three cases of cartridges, it that what you would like to refill?

B: Yes.

A: Is there the correct billing address?

B: No, please post the bill to 124 Hydroid Lane, Milton County, 98830.

A: I will send you an invoice in the next few days. Your order should be delivered before the end of the day on Monday.

B: Thank you.

## Practice

Paul and Leslie are representatives from A Company and B Company. A Company is interested in buying some computer speakers from B Company. Imagine you and your partner are Paul and Leslie, and then make a dialogue with the given points:

- Greet to each other.
- Leslie: make a brief introduction to the product.
- Paul: state clearly about what he wants to buy.
- Make an order with details of quality, quantity, price, packing, delivery, payment, receipt and so on.

## Tips

### ⊠ 下订单常用句型

1. We have pleasure in sending you an order for…

2. We want the goods to be of exactly the same quality as that of those you previously supplied us.

3. Please supply … in accordance with the detail in our order No…

4. This is a trial order. Please send us 50 sets only so that we may tap the market. If successful, we will give you large orders in the future.

5. This order must be filled within five weeks; otherwise we will have to cancel the order.

6. We hope our products will satisfy you and that you will let us have the chance of serving you again.

7. We are pleased to/would like to place an order with you for …

8. We would appreciate it if you could ship these products to us as soon as possible.

9. We hope to receive this order by the end of the month, if possible.

10. I have attached a completed order from this product, as requested.

11. Please bill me using the credit card information on file from my last order.

12. I will send a certified check for USD 1,025.50 by express mail this afternoon.

# WRITING

## How to Write Letter of Complaint and Letter of Apology

## 如何写投诉信和道歉信

### 1. 投诉信(Letter of Complaint)

在人们的生活中经常会发生一些事情，比如消费利益受损、正常生活和工作受到干扰等令人伤脑筋的事情，此时写封投诉信不失为一个解决的办法。

投诉信通常包括以下几个方面的内容：说明投诉的原因并表示遗憾；实事求是地阐述问题发生的经过，切记不要夸大其词；指出问题引起的后果；提出批评及处理的意见或敦促对方采取措施或者提出所希望的赔偿以及补救的方式。

### 2. 道歉信(Letter of Apology)

在工作或生活中，我们经常会遇到突然有事要找同事或朋友帮忙，但是他们却碰巧有事，无法提供帮助，甚至根本联系不到。当然，也有可能正好相反，你的同事或朋友找你有事，你却由于种种原因"辜负"了他们的期望。这时，一封道歉信就必不可少了。

此类信件的写法也比较有规律。一般开头段就要开门见山地表达歉意。第二段具体说明自己没能及时提供帮助的原因，也可以适当地询问对方的问题怎么样了，以及自己现在还能做什么等。第三段则是再次道歉，以示真诚。

## Example

Dear Manager,

I venture to write to complain about the quality of the digital camera I bought last Friday at your store. During the five days the camera has been in my possession, problems have emerged one after another. For one thing, the screen is always black, making the camera no different from a traditional one. For another, the battery is distressing as it supports the camera's operation for only two hours. Therefore, I wish to exchange it for another camera or declare a refund. I will appreciate it if my problem receives due attention.

Sincerely yours,

Li Ming

✉ 尊敬的经理：

　　我冒昧给您写这封信，投诉我上周五在您商店里购买的数码相机的质量问题。购买至今的五天中，各种问题一个接一个地出现。一方面，相机屏幕总是黑屏，我们使用起来仿佛和传统相机没有区别。另一方面，电池的状况也不尽如人意，只能支持相机工作两个小时。因此，我希望可以换一台相机，否则我要求退款。我希望这个问题能够得到足够的重视。

　　　　　　　　　　　　　　　　　　　　　　　　　　　　　　您诚挚的

　　　　　　　　　　　　　　　　　　　　　　　　　　　　　　李明

---

Dear Anne,

Thank you for your invitation to dinner at your home tomorrow evening. Unfortunately, it is much to my regret that I cannot join you and your family, because I will be fully occupied then for an important exam coming the day after tomorrow. I feel terribly sorry for missing the chance of such a happy get-together, and I hope that all of you enjoy a good time. Is it possible for you and me to have a private meeting afterward? If so, please don't hesitate to drop me a line about your preferable date. I do long for a pleasant chat with you.

Please allow me to say sorry again.

Yours truly,

Li Ming

---

✉ 亲爱的安：

　　感谢你邀请我于明日晚上与您和您的家人共进午餐。可是，我非常遗憾地告诉您我无法赴约，因为我将忙于准备后天的一门重要考试。错过了这么一个欢乐的聚会我深感遗憾，我希望你们能度过一个愉快的时光。对了，在我考试后我们可以见一面么？如果可以的话请随时给我打电话，我非常期待能和您愉快地聊天。

　　请允许我再一次致歉。

　　　　　　　　　　　　　　　　　　　　　　　　　　　　　　您真诚的

　　　　　　　　　　　　　　　　　　　　　　　　　　　　　　李明

# Practice

Dear Sir,

_____the deplorable attitude of one of your staff member. _____:
the screen is always black when I do nothing.

However, when I called your Complaints Department, _____
_____. For one thing she interrupted me continually, for another she even said that the fault was my own. _____
_____.


I would like to suggest that the girl in question should be disciplined, and instructed on the proper way to deal with clients. _____.
_____.

Sincerely yours,

Ben

---

✉ 亲爱的先生:

我写这封信是为了让你知道你们一位员工的服务态度。自从上个月从你们那里买了电脑之后,电脑好像有些故障,屏幕经常无故黑屏。

但当我打电话询问客服部门时,接电话的女孩态度非常粗鲁。一方面她总是打断我的话,另一方面她甚至说错误在我。不得不说一下,这样对待客户的态度实在让人难以接受。

我建议对这个女孩进行惩罚,并告诉她应该如何对待客户。我希望她能向我正式道歉。

请早日回复。

您真挚的

本

Dear Prof. Patent,

_____e book report you assigned last week, due to a sudden illness falling upon me a few days ago. _____ _____, which has thus prevented me from any academic activity. I hereby submit the doctor's note.

_____, as my health is turning better.

_____.

Yours sincerely,

Li Ming

✉ 尊敬的 Patent 教授：

我非常遗憾地告知您我没有完成您上周布置的读书报告，因为几天前我突然生病了。过去的几天中我一直持续高烧，住在医院，因此无法进行任何学术活动。随信是医生的诊断书。

如果您能再给一周时间让我来完成这项任务我将感激不尽，因为现在我的身体正在好转。

希望您能理解我的处境并接受我的道歉。

您真诚的

李明

# Tips

✉ 道歉常用句型：

1. Thank you for your invitation to… Unfortunately, it is much to my regret that I cannot…

2. I feel terribly sorry for missing the chance of…

3. I must have caused you a lot of trouble.

4. Please allow me to say sorry again. I am very sorry to inform you that...due to...

5. It was my fault/my mistake. It won't happen again.

6. Hope you can understand my situation and accept my apology.

## ☒ 投诉常用句型：

### 文章开头

1. I am writing to complain about a serious defect found in the computer I bought from you.

2. I am writing to notify you of a claim for damaged goods against your company.

3. I am writing to request you to take corrective actions concerning…

4. I am writing to express my concern/dissatisfaction about …

5. I would like to draw you attention to …

6. It has come to my attention that …

7. I sincerely regret having to write this letter …

8. I would be grateful if you could manage to help me out of the problem.

### 文章末尾：

1. I believe this is the only way to get this matter settled.

2. It is too bad this unfortunate accident occurred. Otherwise, I was very pleased with your service.

3. I am glad to see what you can do to rectify this situation.

4. I urge you to reconsider your consideration.

5. I would be grateful if you would ensure that the same thing does not happen again.

6. If you could find time to let us know, it would set our minds at rest.

7. I am reluctant to take the matter up elsewhere and hope that you will be able to let me have some explanation of the incident.

8. I hope we can work something out to our mutual benefit.

9. Your agreement to the suggested course of action will be appreciated.

10. Your comments will be appreciated.

11. I would like to discuss this matter with you further. Please contact me at 021-047564. I look forward to resolving this with you.

12. I believe this matter can be resolved quickly and look forward to your early reply.

13. Please let me know what you propose in relation to this issue as soon as possible.

14. I am sorry to write to you in this manner and I hope that you will not be offended.

15. If this matter is not put right soon, I fear it could have serious consequences.

# UNIT 8

## TOPICS

- The history of Information Security
- The concept of Information Security
- The goal of Information Security
- What is confidentiality of Information Security?
- What is integrity of Information Security?
- What does availability of Information Security mean?
- Do you know how to make it secure when shopping online?
- What is computer virus?
- How to protect your personal computer from viruses?
- What does firewall technology mean?
- Is it secure after install a firewall?
- Skills of entertaining clients.
- How to write letters of appreciate and congratulation?

 TEXT

# Information Security

Computer networks did not exist at the dawn of the computer Era. Thus, at this time, the need for information security—that is, the need to secure the physical computer and its media, make sure the equipment was not stolen, damaged, or modified since each mainframe was an **isolated** machine. During the 1960s, many more mainframes were brought online. The work performed by these computers started to become time-critical and began to require the use of data and processes **housed** at separate sites. Then, security was becoming a popular concern.

isolate ['aisəleit] vt. 使隔离，使孤立  vi. 隔离，孤立

housed[haust] adj. 封装的

At the close of the twentieth century, the Internet brings millions of computer networks into communication with each other. The personal computer appears on people's desktops at work and even at home. What emerged was a **labyrinth** of networks **boasting** various degrees of security attempting to access and share information with each other. The weaknesses became easier to exploit as access to the Internet's resources became easier.

labyrinth ['læbə,rɪnθ] n. 迷宫；难解的事件
boast [bəust] vt. 自吹自擂

Understanding the aspects of information security requires the definitions of certain information technology terms. The word security means "the quality or state of being secure — to be free from danger". While, the term information security which is referred to computer security and information assurance means protecting information from **unauthorized** access, illegal use, **disclosure**, disruption, modification, **perusal**, **inspection**, recording or destruction.

unauthorized [ʌn'ɔ:θə,raɪzd] adj. 未经授权的；未经许可的
disclosure [dɪ'skləʊʒə] n. (发明等的)公开；泄露，揭晓
perusal [pə'ru:zəl] n. 熟读
inspection [ɪn'spekʃən] n. 检查；检验；视察；检阅

Information security is often achieved by means of several strategies and methods such as polices, standards, and strategies. But the core aim of information security is to ensure the **confidentiality**, **integrity** and availability (CIA) of critical systems and confidential information. CIA is a widely used **benchmark** for evaluation of information systems security, focusing on the three core goals of confidentiality, integrity and availability of information.

confidentiality [kɔnfi,denʃi'æləti] n. 机密性

integrity [in'tegriti] n. 正直，诚实；完整

benchmark ['bentʃ,mɑːk] n. 基准，参照

## Confidentiality

In information security, confidentiality means secrecy of information. The aim of confidentiality is to ensure protection against unauthorized access to or use of confidential information. Consider, for example, credit cards transactions on the Internet are popular with the age of e-trade coming. It requires the credit card number to be transmitted from the buyer to the merchant and from the merchant to a transaction processing network. The system attempts to enforce confidentiality of the card number during transmission by **encrypting** the card information. It is said the confidentiality has occurred if an unauthorized party obtains the card number in any way during the transaction.

encrypt [in'kript] vt.& vi. 把……加密，将……译成密码

To protect confidentiality of information, a number of measures may be used, including:

➢ Information classification

➢ Secure document storage

➢ Application of general security polices

➢ Education of information **custodians** and end users

custodian [kʌ'stəʊdiːən] n. 监护人；管理人；保管人

## Integrity

The second component of information security is integrity. The information created and stored by an organization needs to be stable constantly. Now the word integrity is the term that means data cannot be modified undetectably. That is to assure the data being accessed or read has neither been **tampered** with, nor been altered or damaged through a system error, since the time of the last authorized access.

tamper ['tæmpə] vt. 篡改

When information is exposed to corruption, damage, destruction, or other disruption of its authentic state, that is to say the information is lack of integrity. For example, data stored on disk are expected to be stable and not supposed to be changed. Integrity means that changes should be done only by authorized users and through authorized mechanisms.

## Availability

In general, the availability of CIA is to ensure that information and vital services are assessable for use when required. For any system, the information must be available when it is needed. This means that the computing system which is used to store and process the information must be accessible. For example, research libraries that require identification before entrance. Librarians protect the contents of the library so that they are available only to authorized users. High availability systems aim to remain available at all times, preventing service disruptions due to power outages, hardware failures, and system upgrades. Ensuring availability also involves preventing denial-of-service attacks.

Availability, like other aspects of security, may be affected by purely technical issues, natural **phenomena**, or human causes.

phenomena [fi'nɔminə] n. 现象

Information security is a "well-informed sense of assurance that the information risks and controls are in balance." Information Security is usually achieved through a mix of technical, organizational and legal measures. These may include the application of **cryptography**, the **hierarchical** modeling of organizations in order to assure confidentiality, or the distribution of accountability and responsibility by law, among interested parties.

cryptography [krip'tɔgrəfi] n. 密码使用法，密码系统，密码术
hierarchical [ˌhaiə'rɑːkikəl] adj. 按等级划分的，等级(制度)的；分层的

 **EXERCISES**

Ⅰ. Match the terms and interpretations.

1. Security Awareness
2. Confidentiality
3. Integrity
4. Availability
5. Encryption

(a) A measure of the degree of a system which is in the operable and committable state at the start of mission when the mission is called for at an unknown random point in time.

(b) The conversion of data into a form that cannot be easily understood by unauthorized people.

(c) The knowledge, skill and attitude an individual possesses regarding the protection of information assets.

(d) The characteristic that alterations to a system's assets can be made only in an authorized way.

(e) Refers to the property of a computer system whereby its information is disclosed only to authorized parties.

Ⅱ. Are the following statements True (T) or False (F)?

1. ( ) The Internet is more secure since more people are using it.
2. ( ) Changing passwords on your computer, and using combinations of letters and numbers, makes it easier to gain access.
3. ( ) Information security systems need to be implemented to protect personal staff details in many businesses.
4. ( ) Computer viruses are programs that can infect other computer programs.
5. ( ) Computer security concerns only with the software and data of computer systems.
6. ( ) Data backup means making copies of data.
7. ( ) Hacker means a bad man who accesses to a computer system without authority.
8. ( ) A backup file is a copy of a file which is kept in case anything happens to the original file.
9. ( ) Data is said to be corrupt only when lost.
10. ( )In computer security, access control has been widely used to manage the permissions

of users to access the objects.

III. Translate the following words and phrases into Chinese.

1. Computer Viruses　　　　　　　　　＿＿＿＿＿＿＿＿＿＿＿＿＿＿＿
2. Network Worms　　　　　　　　　　＿＿＿＿＿＿＿＿＿＿＿＿＿＿＿
3. Access Control　　　　　　　　　　＿＿＿＿＿＿＿＿＿＿＿＿＿＿＿
4. Physical Security　　　　　　　　　＿＿＿＿＿＿＿＿＿＿＿＿＿＿＿
5. Firewall　　　　　　　　　　　　　＿＿＿＿＿＿＿＿＿＿＿＿＿＿＿

IV. Translate the following Chinese statements into English.

1. 互联网使用得越多，对互联网安全就关注得越多。
_____

2. 有效的信息安全包括了一系列政策、安全产品、技术等。
_____

3. 对于多数公司而言，80%的安全问题来自于公司内部。
_____

4. 黑客有两种攻击计算机系统的方法。
_____

5. 一个有效的保护计算机网络的方法是安装防火墙。
_____

V. Fill in each of the blanks with one of the following words or phrases.

　　*refer to*　　*transmitted*　　*collection*　　*crime*　　*information*
　　*recorded*　　*valuable*　　*disrupt*　　*permission*　　*data*

Any part of a computing system can be the target of a_____. When we_____a computing system, we mean a_____of hardware, software, storage media,_____, and people that an organization uses to perform computing tasks. Sometimes, we assume that parts of a computing system are not_____to an outsider, but often we are mistaken. For instance, we tend to think that the most valuable property in a bank is the cash, gold, or silver in the vault. But in fact the customer_____in the bank's computer may be far more valuable. Stored on paper,_____on a storage medium, resident in memory, or_____over telephone lines or satellite links, this information can be used in myriad ways to make money illicitly. A competing bank can use this information to steal clients or even to _____ service and discredit the bank. An unscrupulous individual could move money from one account to another without the owner's _____. A group of con artists could contact large depositors and convince them to invest in fraudulent schemes. The variety of targets and attacks makes computer security very difficult.

Ⅵ. The following is a conversation between customer and tech support. Fill in the blanks with the help of the Chinese given in the brackets.

Tech Support:   What can I do?

Benjamin:   _____ (我在网上冲浪时电脑系统崩溃了)

Tech Support:   _____ (你浏览过非法网站吗？)

Benjamin:   No, but does that matter?

Tech Support: Yes, _____ (如果浏览非法网站，电脑很容易感染病毒)

Benjamin:   I see. I'd better never try.

Tech Support:   That's wise.

Benjamin:   Do you know what's wrong with my PC?

Tech Support: One minute. _____ (哦，它中毒了，而且电脑上没有杀毒软件)

Benjamin:   Is the software necessary for a PC?

Tech Support:   Of course. You'd better learn something about it.

Benjamin:   I'm afraid yes. But what about the data I stored in the computer?

Tech Support:   Don't worry. It should have been protected automatically. _____ (我有杀毒软件，需要我帮你安装吗？)

Benjamin:   Yes, please. I'll really appreciate that.

# SUPPLEMENTARY

## Viruses Can Eat Your Computer Alive

System management is the day to day running of a computer system. Any computer system that is doing useful work must be maintained on a regular basis. But each year, the number of security vulnerabilities discovered rises, and hacking tools available to exploit these vulnerabilities become more readily available and easier to use. Hackers look for ways to "access your computer and copy, steal or alter data you have stored". Absolutely and without question, the number one computing- and networking-related challenge we all face today is computer and network security.

Computer and network security problems affect corporations and small businesses, governments, higher education, schools, individuals and families alike—all of us are struggling.

No matter how careful we might be, computers may be found infected with different viruses. If using Internet frequently for downloading movies, music and other files from the Internet, the chances of picking up a trojan, worm, or other virus are almost assured.

A virus invades for system, erases the hard drive and sends random documents to everyone on the user's contacts list. A computer virus is a typically malicious program written to spread itself easily from computer to computer. Its sole purpose is to infect, corrupt and ultimately breakdown a computer system and any network it may be a part of.

A computer virus has few boundaries and limitations, and with the global use of E-mail and the internet allowing a virus to spread so easily, a simple click of the mouse can cause almost instant worldwide chaos. Without prior knowledge of any possible virus that is circulating at the time, this type of computer virus is impossible to spot. The user will have no idea that the mail that they have just clicked on and opened has infected their PC and, in some cases, the PC virus will have already sent itself out to all of the email addresses contained within the users email contact list. There are some most common viruses:

> **The Panda Burns Incense computer worm**—this was virus wreaked havoc for months in China in 2006 and 2007. Jumping one computer to another by tricking users into opening what appeared to be a friendly E-mail message, the Panda funneled passwords, financial information and online cash balances leaving a panda as its calling card.

> **The Melissa Virus**—this was a bug that hit every PC users. It would automatically E-mail itself to other people without permission. It can be extra harmful if using a private mail server. The Melissa virus has gone down in history as one of the most common computer viruses of all time.

> **A Trojan horse**, or Trojan, is a standalone malicious program designed to give full control of infected PC to other PC. It can also perform other typical computer virus activities. Trojan horses can make copies of themselves, steal information, or harm their host computer systems. The term is derived from the Trojan Horse story in Greek mythology

because Trojan horses attempt to give full control of PC to other PC.

Thankfully, there are many great virus protection programs on the market today that can instantly vanquish even the toughest viruses, but that doesn't mean the user shouldn't have an idea of some of the common computer viruses that are currently going around.

When using computers especially surfing on the Internet, there are some tips to follow:

1. Don't click on pop-up ads that advertise antivirus or anti-spyware programs. If you are interested in a security product, contact the retailer directly through its home page, retail outlet or other legitimate contact methods.

2. Don't download software from unknown sources. Some free software applications may come bundled with other programs, including malware.

3. Use and regularly update firewalls, antivirus, and anti-spyware programs. Keep these programs updated regularly. Use the auto-update feature if available.

4. Patch operating systems, browsers, and other software programs. Keep your system and programs updated and patched so that your computer will not be exposed to known vulnerabilities and attacks.

5. Regularly scan and clean your computer. Scan your computer with your anti-spyware once a week.

6. Back up your critical files. Back-up copies of important files is one of the most fundamental requirements if the computer becomes infected.

# CONVERSATION

## Entertaining Clients

## 接 待 客 户

在经济日益发达的今天，人与人之间的距离逐渐缩短。在日常工作及社会交往中，接待客户的活动越来越多。但是如何接待客户对每个人而言都是一件值得学习的事。下面是在日常接待中应该注意的事项。

1. 迎接礼仪

迎来送往，是社会交往接待活动中最基本的环节，是表达主人情谊、体现礼貌素养的重要方面。尤其是迎接，能给客人留下重要的第一印象。通常迎接客人要有周密的部署，应该重点注意以下事项：
- 了解客人的车次、航班。安排与客人身份、职务相当的人员前去迎接。
- 迎接客人时应该提前为客人准备好交通工具。将客人送往目的地后，应对当地的风土人情、自然景观等做简单介绍。

2. 接待礼仪

接待客人要注意以下几点：
- 客人要找的负责人不在时，要明确告诉对方负责人何时回来。
- 客人到来时，如果负责人不能马上接见，要向客人说明等待理由与等待时间。
- 接待人员带领客人到达目的地，应该有正确的引导方法和引导姿势。

# Example

## Dialogue 1

A: How can I help you?

B: Yes, I am Jim Milson from Edison associate. I'd like to see Mr. Smith.

A: Do you have an appointment?

B: Yes, he knows I am coming. Our meeting is at 9 o'clock.

A: I wondering if Mr. Smith forgot your meeting. I am afraid he left this office this morning and he is not expected back until after 4 PM. Let me find out if he made an arrangement for someone else to meet with you in his place. Will you please have a seat?

B: Sure.

A: Yeah, Mr. Milson. I just checked with their office manager Ms Terry, she said Mr. Smith briefed her on your project. She is just finishing up our meeting now. She should be meeting you shortly. Would you like me to show you around for your waiting?

B: That would be very nice, thank you!

A: Right this way Mr. Milson. We can start with our front office. When Ms Terry is ready, you may take at the front 6th floor. There is a conference room already prepared.

## Dialogue 2

F: Hello, Mr. Henson, welcome to Beijing! Is this your first time to visit China?

M: Oh no, I've already made several trips to Guangzhou, this is my first trip to Beijing. It is a lot larger than I expected it would be.

F: Yes, Beijing has been broken over the last few years, there are a lot of improvement changing be made for Olympic. What would you like to see when are you here?

M: I hope to have time to visit Great Wall when I am here. I always want to go there. I think it would be a real shame by came all the way in Beijing and didn't make out the wall. Do you think I have a chance to see it?

F: I can pretty sure it can be arranged. The Wall is a short distance from the city. But we could make arrangements for driver to take us out to visit the Great Wall during when our afternoon breaks. I also recommend you to visit Tian'an Men Square and city while you add it!

M: Yes, that would be nice. Would I have a tour guide to tour completely visit these places?

F: Don't worry. I would be able to go along with you over the next few days. If you have any questions or problems I will be right here to help you out. I can be a translator and tour guide.

M: Thank you very much.

F: My pleasure. I hope your visit to Beijing is very enjoyable!

## Practice

Imagine that A is a sales manager of ABC Company, and B is a guest of ABC. A is going to pick B up at the airport. Please complete the following dialogue between A and B with the given points:

- ⋄ Greet each other.
- ⋄ Talking about the trip.
- ⋄ Talking about the hotel.
- ⋄ Make a brief introduction to the city and local customs.

## Tips

### 招待客户常用语

1. I'm XXX, from AA. I've come to meet you. 我是从 AA 来的 XXX，我是来接您的。

2. Did you have a pleasant trip?/How was your trip? 旅途愉快吗？

3. Is this the first time you com to Shenzhen? 您是第一次来深圳吗？

4. The car is in the parking lot, this way please. 车在停车场，这边请。

5. Would you like a glass of water? /Can I get you a cup of Chinese red tea? / How about a Coke? 来杯水怎么样？/我给您倒杯红茶吧？/来杯可乐怎么样？

6. A cup of coffee would be great. Thanks！来杯咖啡吧。谢谢！

7. There are many places where we can eat. How about Cantonese food? 有很多吃饭的地方。粤菜怎么样？

8. I would like to invite you for lunch. 我想请您吃午饭。

9. Oh, I can't let you pay. It is my treat, you are my guest. 哦，不能让您请。这次我来，您是客人。

10. May I propose that we break for coffee now? 我们休息一下喝杯咖啡吧？

11. Excuse me. I'll be right back. 不好意思，我马上回来。

12. Have a good journey！祝您旅途愉快！

13. Thank you very much for everything you have done us during our stay in China. 谢谢您为我们在中国期间所做的一切。

14. I'm looking forward to seeing you again. 希望能再次见到您。

15. There is a welcoming dinner tomorrow night. 明晚将有一个接风宴。

16. I will take you on a city tour tomorrow morning. 明天上午我会带您浏览市容。

17. Would you please tell me when you are free? 能告诉我您什么时候有空吗？

18. I'm afraid I have to cancel my appointment. 恐怕我得取消约会了。

19. Will you change our appointment tomorrow at 10:00 a.m. to the day after tomorrow at the same time? 我们的约会是不是要从明天上午10点改到后天同一时间？

20. Would you like to go through our factory some time? 什么时候来参观我们的工厂吧？

21. Thank you for coming today. 谢谢您的莅临。

22. Wouldn't you like to spend an extra day or two here? 您不愿意在这里多待一两天吗？

23. When can we work out a deal? 我们什么时候洽谈生意？

24. I'd appreciate your kind consideration in the coming days. 在接下来的几天里还请多关照。

 WRITING

## How to Write Effective Letters of Appreciation and Congratulation

## 如何写感谢信及祝贺信

在日常生活和社会交往中，我们常常会得到别人的帮助、招待、馈赠、慰问等。而感谢信作为一种礼仪文书，主要用来向对方表示出自然且充满真挚的感谢之情。感谢信在写作时通常篇幅较短，直接表达对对方的感谢之情。

在英语写作中，祝贺信使用的频率很高。在同学、同事、朋友或者亲戚取得成绩、获得奖励以及结婚、生子、生日、晋升等情况下，通常要写祝贺信以表达我们的祝贺之情。祝贺信用词必须亲切有礼、表达出真诚的喜悦之情。

英语感谢信和祝贺信的写作格式类似，主要由以下几部分组成：

➢ 信头(heading)，即发信人的地址和日期(右上角)；
➢ 称呼(salutation)，即对收信人的尊称(一般用 Dear Mr…, Dear Madam…, Dear Miss…等)；
➢ 正文(body)，即感谢信或祝贺信的主要内容，通常正文第一句和称呼之间要空一至二行；
➢ 信尾客套话(complimentary close)或祝贺词，即写信人在信的右(或左)下角写上表示自己对收信人的一种礼貌客气的谦称或者再次表示感谢或祝贺。

## Example

December 22, 2011

Dear Mrs. Beck,

How did you ever find those wonderful glasses? They are perfect, and Jim and I want to thank you a thousand times!

The presents will be shown on the day of the wedding, but do come over Thursday morning if you can for a cup of coffee.

Thank you again, and with love from us both.

Mary

亲爱的贝克太太：

您是怎样弄到这些漂亮的玻璃杯子的？它们漂亮极了。吉姆和我向您表示万分的感谢！举行婚礼那天，所有的礼品都要展示出来。如果可能，请周二早上一定过来喝杯咖啡。再次表示感谢，并致以我们俩的问候。

玛丽
2011 年 12 月 22 日

---

Dear Jane,

Congratulations, DR. Jane. I love the way that sounds! Now you're graduating with high honors from your University. We all proud of you!

I wish you greater success and fulfillment in the years ahead.

With best wishes

Sincerely yours,

Jack

---

✉ 亲爱的詹妮：

恭喜你，詹妮博士！我喜欢这样称呼你。现在你已经以优异的成绩毕业了。这真是太好了！我们都以你为荣！

希望你在以后的日子里取得更大的成绩！

谨致良好的祝愿！

您诚挚的
杰克

## Practice

Dear Mr./Ms,

_____

_____

_____

Yours faithfully,

_____

▷ 尊敬的先生/小姐：

非常感谢上周四您能抽出时间和我讨论新项目的进展。您对该项目的观点让我受益匪浅，希望我们能有进一步的合作。

再次表示感谢，并致以诚挚的问候。

您诚挚的

XXX

Dear Mr. Hills,

_____

_____

_____

Yours faithfully,

Tom

▷ 亲爱的 Hills 先生：
至此 2012 年春节即将到来之际，我谨向您致以最良好的祝愿和问候。
希望您能在中国度过愉快的一年。

您诚挚的
汤姆

# Tips

▷ 感谢信及祝贺信常用语

1. Thank you for your support. 谢谢您的支持。

2. Thank you for your hospitality that made us feel as comfortable as we do in our own home. 谢谢您的盛情款待，让我们感觉就像在自己家里一样。

3. Many thanks for your kind and warm help. 谢谢您的帮助。

4. I truly appreciate your kind letter and show our gratitude for your good wishes. 非常感谢您的亲切来函，并对您的祝福深表谢意。

5. We are indebted for … 我们非常感谢……

6. It was really exciting to get your New Year's card! 很高兴收到您的新年贺卡。

7. I'm sincerely grateful for all your help. 非常感谢您的帮助。

8. Your note of congratulations is deeply appreciated. 非常感谢您的祝福。

9. Thank you for doing so much to make my trip interesting. 谢谢您让我的旅途如此愉快。

10. Thank you for your kindness to have done me a favor. 谢谢您对我的帮助。

11. Heartfelt congratulation on… 忠心地祝福……

12. I am so happy to hear that… 我很高兴地听到……

13. I write to congratulate on… 我写信来祝贺……

14. I offer you my warmest congratulations on your… 向您的……表示祝福。

15. I wish you still further success! 希望您能取得进一步的成功！

16. I was pleasantly surprised to hear that you… 我惊喜地获悉……

17. Please accept our sincerest congratulation. 请接受我们最真诚的祝福。

18. Congratulations to you on being award the scholarship. 祝贺您获得奖学金。

19. May lasting happiness forever! 祝您永远幸福！

20. It is wonderful to know that you … 知道您……我很高兴。

# UNIT 9

## TOPICS

- What is multimedia?
- What are the basic elements of multimedia on a computer?
- What are the requirements of a multimedia PC?
- What does multimedia do?
- What do you think of the future of multimedia?
- What is the difference between vector and bitmap graphics?
- Do you know about some image/video editing software?
- Business negotiation skills.
- How to write a notice or an invitation letter?

 TEXT

# What Is Multimedia?

Multimedia is simply multiple forms of media integrated together. Media can be text, graphics, audio, **animation**, video, data, etc. An example of multimedia is a web page on the topic of **Mozart** that has text regarding the **composer** along with an audio file of some of his music and can even include a video of his music being played in a hall.

Multimedia itself has its binary aspects. As with all modern technologies, it is made from a mix of hardware and software, machine and ideas. More importantly, you can **conceptually** divide technology and function of multimedia into control systems and information. The enabling force behind multimedia is digital technology. Multimedia represents the **convergence** of digital control and digital media—the PC as the digital control system and the digital media being today's most advanced form of audio and video storage and **transmission**. In fact, some people see multimedia simply as the marriage of PCs and video. PC power has reached a level close to that needed for procession television and sound data streams in real time, multimedia was born. Multimedia PC needs to be more powerful than mainstream computer—at least the multimedia PC defines the mainstream. Among **contemporary** PCs, about the only things that separate an ordinary computer from multimedia are a soundboard and a CD-ROM driver. The CD serves as multimedia's chief storage and exchange medium. Without the

animation [ˌæniˈmeiʃən] n. 活泼, 生气, 兴奋, 动画片, 动画片制作
Mozart n. 莫扎特(人名)
composer [kəmˈpəuzə] n. 创作者(尤指乐曲的)

conceptually [kənˈseptjuəli; kənˈseptʃuːəli] adv. 概念地

convergence [kənˈvəːdʒəns] n. 收敛, 汇聚, 汇合点

transmission [trænsˈmiʃən] n. 传输, 传播, 播送, 变速器

contemporary [kənˈtempə.reri] adj. 同时代的, 当代的 n. 同时代的人, 同龄人

convenient CD, the PC industry would lack a means of **distributing** the hundreds of megabytes of audio, visual, and textual data that make up today's multimedia titles. Without CD, you couldn't buy multimedia because publishers have no way of getting it to you.

distributing [dis'tribju(:)tiŋ] adj. 分布的；及物动词 distribute 的现在分词

So what is multimedia? By now you should agree that multimedia isn't any one thing but a complex entity that involves the many things: hardware, software, and the interface where they meet. But we've forgotten the most important thing that multimedia involves: you. Yeah, sure. With multimedia, you don't have to be a **passive** recipient. You can control. You can **interact**. You can make it do what you want it to do. It means you can **tailor** a multimedia presentation to your own needs. You can cut through the chaff and dig directly into the important data in a report, pull together reports and video clips from around the world that interest you. That's the strength of multimedia and what distinguishes it from traditional media like books and television.

passive ['pæsiv] adj. 被动的, 消极的 n. 被动性

interact [,intə'rækt] vi. 相互作用, 相互联系, 相互影响, 互动

tailor ['teilə] n. 裁缝 vt. 缝制, 调整使适合 vi. 做裁缝

What does multimedia do? Now we advocated one thing. The thing is that use multimedia to expand the uses of computers. Let's take a look at some certain areas. Multimedia could have a direct impact to these areas:

(1) **Computer-based training (CBT)**

Many companies are turning to multimedia applications to train their employees. They found it has saved expenses and trained employees more effectively by using a multimedia application.

(2) **Education**

The essence of multimedia is to make computers more interesting. Multimedia can make the learning process more interesting. So it will help the learning process.

(3) **Entertainment**

In many cases, today's best games use the graphics technology. In addition, writing entertainment applications (that is, games) can be a lot of fun.

(4) **Information access**

This is the age of information. We often find nothing with so much information. Multimedia provides effective ways to organize information and search for facts.

(5) **Business presentations**

To many companies, presenting information to business professionals is a required form of communication. Applications are already available for creating great-looking presentations, and through multimedia these applications will become even better and more effective.

Obviously, multimedia has many uses, and the only limitation is your imagination.

As technology progresses, so will multimedia. Today, there are plenty of new media technologies being used to create the complete multimedia experience. For instance, **virtual reality** integrates the sense of touch with video and audio media to **immerse** an individual into a virtual world. Other media technologies being developed include the sense of smell that can be transmitted via the Internet from one individual to another. Today's video games include **bio feedback**. In this instance, a shock or **vibration** is given to the game player when he or she crashes or gets killed in the game. In addition, as computers increase their power, new ways of integrating media will make the multimedia experience extremely **intricate** and exciting.

virtual reality 虚拟现实
immerse [i'mə:s] vt. 浸, 陷入

bio feedback 生物反馈
　bio 为 biology (生物学)的缩写
vibration [vai'breiʃən] n. 振动, 颤动

intricate ['intrəkit] adj. 错综复杂的

# EXERCISES

I. Match the terms and the interpretations.

1. Multimedia PC (MPC)

2. Computer-Based Training (CBT)

3. Vector Graphics

4. Adobe Premiere Pro

5. Virtual reality (VR)

(a) A real-time, timeline based proprietary video editing software application. It is part of the Adobe Creative Suite, a suite of graphic design, video editing, and web development applications developed by Adobe Systems, though it can also be purchased separately.

(b) A recommended configuration for a PC with a soundboard and a CD-ROM drive. The standard was set and named by the "Multimedia PC Marketing Council", which was a working group of the Software Publishers Association (now the Software and Information Industry Association).

(c) A term that applies to computer-simulated environments that can simulate physical presence in places in the real world, as well as in imaginary worlds.

(d) The use of geometrical primitives such as points, lines, curves, and shapes or polygon(s), which are all based on mathematical equations, to represent images in computer graphics.

(e) Synonyms to e-learning. The information and communication systems, whether networked learning or not, serve as specific media to implement the learning process.

II. Are the following statements True (T) or False (F)?

1. (    ) Multimedia file formats are capable of storing sound and video information.

2. (    ) Vector files contain data described as mathematical equations.

3. (    ) Bitmapped images often have a cartoon-like appearance.

4. (    ) Vector graphics are stored as a vertical and horizontal array of pixels.

5. (    ) Bitmapped images are made up a series of pixels in a grid.

6. (    ) Victories drawings can be enlarged as much as desired.

7. (    ) Bitmap pictures can not be enlarged as much as desired.

8. (    ) Video movies are large before and after rendering, and you need a lot of space to save them.

9. (    ) Bitmapped images are resolution independent.

10. (    ) Top video editing software is a pricey purchase.

III. Translate the following words and phrases into Chinese.

1. Virtual Reality                               _____
2. Learning Curve                           _____
3. Streamlined                                 _____
4. Response Time                           _____
5. Workflow                                     _____
6. Video Editing                             _____
7. VCR                                             _____
8. Industry Standard                     _____
9. Blu-Ray                                       _____
10. Macintosh (Mac)                       _____

IV. Translate the following Chinese statements into English.

1. 多媒体技术需要同时处理声音、文字、图像等多种媒体信息。

2. 矢量图适于描述由多种比较规则的图形元素构成的图形。

3. MIDI是数字乐器接口的国际标准，它定义了电子音乐设备与计算机的通信接口。

4. 视频是连续渐变的静态图像顺次更换显示，从而构成运动视感的媒体。

5. 多媒体信息虽然经过了压缩处理，但还是含有大量的数据，所以需要有大容量的存储设备来保存这些信息。

---

V. Fill in each of the blanks with one of the following words.

*multiple   hypermedia   foreign   performance   page   audio   media   stand for*

*pictures   video   back*

Multimedia is simply _____ forms of media integrated together. Media can be text, graphics, audio, animation, video, data, etc. An example of multimedia is a web_____ on the topic of Mozart that has text regarding the composer along with an _____ file of some of his music and can even include a _____ of his music being played in a hall.

Besides multiple types of _____ being integrated with one another, multimedia can also _____ interactive types of media such as video games, CD-ROMs that teach a _____ language, or an information Kiosk at a subway terminal. Other terms that are sometimes used for multimedia include _____ and rich media.

The term Multimedia is said to date _____ to 1965 and was used to describe a show by the Exploding Plastic Inevitable. The show included a _____ that integrated music, cinema, special lighting and human performance. Today, the word multimedia is used quite frequently, from DVDs to CD-ROMs to even a magazine that includes text and _____.

# SUPPLEMENTARY

## Where Is Virtual Reality?

Have you ever used Virtual Reality? Whether the big cumbersome headsets in the early 90's or the sleeker more refined technology of today, chances are you've encountered virtual reality in your travels. It was touted as the "next big thing" in computers and was expected to be in every home but it never eventuated. Today I'm going to discuss what virtual reality means for gamers and game developers.

## Introduction

I first used virtual reality in 1995 while visiting SeaWorld in London. They had a ride you sat in, donned a virtual reality headset and furiously pressed buttons to win or lose a battle (along with 20 or so other people). I was so excited to finally get to use virtual reality that I took a wrong turn getting on the ride and fell off into the hydraulics. I'm not sure if it was the amazing 3D, head tracking or loss of blood from my injuries but I had an awakening experience. I knew that in the future, we'd all be playing games using virtual reality, and the world would be a better place. Unfortunately, this never happened, but is it too late?

## Key features of Virtual Reality

VR has a number of key features that make it the obvious path for game development in the future:

- **Full 3D Immersion**—rather than looking at the game world sitting on a monitor that's sitting in your room you are actually in the world. Even 2D games take up your entire peripheral vision making them immersive.

- **True 3D (for free)**—Having two images allows a true stereoscopic display of the 3D world.

- **Head tracking**—Most VR headsets have built in head tracking which adds to the immersion. Now you can look at the pre-pubescent kid in counter strike before he headshots you.

## Why hasn't Virtual Reality taken off ?

There are a few main reasons why Virtual Reality is still not in the mainstream of gaming:

- **Price**—With the average headset coming in at $1,500 USD, this is a lot for an 800×600 display by anyone's standards. While it is comparable to some larger end monitors it's still too high for the average gamer. If the price of headsets can come down below $750 USD and resolutions can be at least 1024×768 I expect a much larger uptake. That's certainly the sweet spot for price and resolution for me.

- **Vertigo/Eye Strain**—Some people have issues focusing on the screens in the virtual reality headset and others suffer from eye strain. Most people should get used to this over time however it might simply mean some people never adopt the technology.

- **Lack of interest**—People just got bored with the idea. For so long we kept hearing about how VR was going to change the world and it never did. A great example of this is the fact that searches for "Virtual Reality" are down 80% since 2004!

## What do you think?

Have you used Virtual Reality before? Did you like it? How much would you be willing to pay and what resolution do you think should be the minimum supported? Many people I've spoken to about Virtual Reality think it's dead, what do you think?

# CONVERSATION

## 8 Proverbs for Business Negotiation

## 商务谈判的八字箴言

在商务谈判中，双方谈判能力的强弱差异决定了谈判结果的差别。对于谈判中的每一方来说，谈判能力都来源于八个方面，即 "NO TRICKS" 中每个字母所代表的八个单词：need, options, time, relationships, investment, credibility, knowledge, skills.

● N 代表需求(need)。对于买卖双方来说，谁的需求更强烈一些？如果买方的需要较多，卖方就拥有相对较强的谈判力；如果卖方非常希望卖出他的产品，买方就拥有较强的谈判力。

● O 代表选择(options)。如果谈判不能最后达成协议，那么双方会有什么选择？如果对方认为你的产品或服务是唯一的或者没有太多选择余地，你就拥有较强的谈判资本。

● T 代表时间(time)。谈判中可能出现有时间限制的紧急事件，如果买方受时间的压力，自然会增强卖方的谈判力。

● R 代表关系(relationship)。如果与顾客之间建立了强有力的关系，在同潜在顾客谈判时就会拥有关系力。但是，也许有的顾客觉得卖方只是为了推销，因而不愿建立深入的关系，这样，在谈判过程中将会比较被动。

● I 代表投资(investment)。在谈判过程中投入了多少时间和精力？为此投入越多，对达成协议承诺越多的一方往往拥有较少的谈判力。

● C 代表可信性(credibility)。潜在顾客对产品可信性也是谈判力的一种。如果推销人员知道你曾经使用过某种产品，而他的产品具有价格和质量等方面的优势时，无疑会增强卖方的可信性，但这一点并不能决定最后是否能成交。

● K 代表知识(knowledge)。知识就是力量。如果你充分了解顾客的问题和需求，并预测到你的产品能如何满足顾客的需求，你的知识无疑增强了对顾客的谈判力。反之，如果顾客对产品拥有更多的知识和经验，顾客就有较强的谈判力。

● S 代表的是技能(skill)。这可能是增强谈判力最重要的内容了，不过，谈判技巧是综合的学问，需要广博的知识、雄辩的口才、灵敏的思维等各方面的素质。

## Example

### ⊠ Dialogue 1

A: So, thank you for coming, everyone. It's really a pleasure to see you all here. First of all, may I suggest you take a look at the agenda i sent you? Would you like to make any comment on that?

B: Yes, I wonder if we can begin with shipment question first. We really need to come to an agreement on that before anything else.

A: That's true, but it's also a very difficult issue. That's the reason why I put it last. I thought it might be a good idea for us to start with the points we have in common. We'll move on to the shipment issue after that.

B: All right. That sounds reasonable.

A: Well, before we go any future, I would like to say strongly how I feel that it's in both our interest to reach an agreement today. The market is becoming even more competitive and our combined strength will give us some big advantages, not least in terms of the dealer network. Now, I think Richard would like to say a few words about that.

### ⊠ Dialogue 2

A: George, We have decided the price, now let's get down to some detail requirements of the products you order. First, we'd like to know how you would like the flowers are packed.

B: For No.10, each bunch of flower should be packed in a clear transparent plastic bag, each bag to a paper box, 100 boxes to a carton. We require the plastic bags should be in 7 different colors, and the quality of each bag should be grade AAA with degree of transparency of 100%.

A: Grade AAA is large spend for us, we can't meet your standard. The most we can do is to use grade A. If you insist, we have to take 15 cents extra charge for each bag.

B: If you can guarantee the quality and make sure each bunch of flower to reach customers without defections, I can agree that.

A: Please don't worry, George, we can guarantee.

B: For No.20 and No，30, each pot of flower to a wooden box, each box should be moisture-conditioned. 6 articles of No.20 and 6 articles of No.30 to a wooden case.

> A: It will be a waste of wood, if you use box and case at the same time. we can use wood blocks to separate the flowerpots.
> B: No, I can't agree. If having no case packing, the flowers are inconvenient to convey and easy to be broken. So please do as I say.
> A: Ok.And what's the requirement for shipping marks?
> B: We need cartons and wooden boxes painting our company name for short. And of course, some indication marks, such as fragile, keeping upright should be put on.
> A: Anything more?
> B: I think no more. That's all.

## Practice

Imagine that you and your partners are representatives from AAA Company and BBB Company respectively. BBB Company is interested in importing some scarves made in different materials from AAA Company. Practice the following negotiation skills with the given points:

- ◇ Greet to each other.
- ◇ Make a brief introduction to your company.
- ◇ AAA Company: Make a brief introduction to the product.
- ◇ BBB Company: State your company's requirements about the product.
- ◇ Negotiate about the price, packing, delivery, payment, insurance and so on.
- ◇ Make an agreement.

## Tips

▷ 商务谈判常用语

1. Would anyone like something to drink before we begin? 在我们正式开始前，大家喝点什么吧？

2. I know I can count on you. 我知道我可以相信你。

3. We are here to solve problems. 我们是来解决问题的。

4. We'll come out from this meeting as winners. 这次会谈的结果将是一个双赢。

5. I hope this meeting is productive. 我希望这是一次富有成效的会谈。

6. I need more information. 我需要更多的信息。

7. Not in the long run. 从长远来说并不是这样。

8. Let me explain to you why. 让我给你解释一下原因。

9. That's the basic problem. 这是最基本的问题。

10. Let's compromise. 让我们还是各退一步吧。

11. It depends on what you want. 那要视贵方的需要而定。

12. The longer we wait, the less likely it is we will come up with anything. 时间拖得越久,我们成功的机会就越少。

13. I'm sure there is some room for negotiation. 我肯定还有商量的余地。

14. We have another plan. 我们还有一个计划。

15. Let's negotiate the price. 让我们来讨论一下价格吧。

16. We could add it to the agenda. 我们可以把它也列入议程。

17. We cannot be sure what you want unless you tell us. 希望您能告诉我们,要不然我们无法确定您想要的是什么。

18. We have done a lot. 我们已经取得了不少进展。

19. We can work out the details next time. 我们可以下次再来解决细节问题。

20. I suggest that we take a break. 我建议咱们休息一下。

21. Let's recess and return in an hour. 咱们休会,一个小时后再继续。

22. We need a break. 我们需要暂停一下。

23. May I suggest that we continue tomorrow? 我建议明天再继续,好吗?

24. We can postpone our meeting until tomorrow. 我们可以把会议延迟到明天。

25. That will eat up a lot of time. 那会耗费很多时间。

WRITING

# How to Write a Notice or an Invitation Letter

# 如何写通知及邀请信

1. 通知(Notice)

通知是上级对下级、组织对成员或平行单位之间部署工作、传达事情或召开会议等所使用的应用文。通知有两种:一种是书信方式,寄出或发送,通知有关人员,此种通知写作形式同普通书信,只要写明通知的具体内容即可;另一种是布告形式,张贴通知。究竟采用哪种形式，视实际需要而定。如通知的对象为集中的较大范围的人员，例如学生、会员、读者、观众等，宜采用布告的形式。这种通知一般在上方居中处写上"Notice"或"NOTICE"一词作为标志，正文的下面靠右下角写出通知的单位名称或人名，也可放在正文之上。有时出通知的单位名称会写在正文的开头，这样就不需要另外注明。出通知的日期写在正文的左下角。如有单位负责人署名，可写在右下角。出通知的单位以及被通知的对象一般都用第三人称。如果正文前用了称呼用语，则用第二人称表示通知对象。通知要求言简意赅、措辞得当、时间及时。

2. 邀请信(Invitation Letter)

邀请信包括宴会、舞会、晚餐、聚会、婚礼等各种邀请信件，形式上大体分为两种：一种为正规的格式(formal correspondence)，亦称请柬；一种是非正式格式(informal correspondence)，即一般的邀请信。邀请信在形式上不如请柬正规，但也很考究。书写时应注意：邀请信一定要将邀请的时间(年、月、日、钟点)、地点、场合写清楚，不能使接信人存在任何疑虑。例如"I'd like you and Bob to come to Luncheon next Friday."这句话中所指的是哪个星期五并不明确，所以应加上具体日期："I'd like you and Bob to come to luncheon next Friday, May the fifth."。

# Example

**NOTICE**

The sports meeting which was to take place this Saturday has to be put off because of the heavy rain these days. All students are required to come to school on Saturday morning as usual, but there will be no class that afternoon. Weather permitting, the sports meeting will be held next Saturday morning. Members of the school ping-pong team must come to the Ping-pong Hall at 4:30 this Saturday afternoon. Ping-pong stars from Wuhan will come and give special training and coaching then.

Office of Physical Education

10th September, 2009

✗                  通 知

　　由于近日连降大雨，我校原定本周六举行的运动会将作延期。同学们周六早晨照常到校上课，周六下午停课。如果天气允许的话，运动会将在下周六早晨举行。学校乒乓球队的队员们请于本周六下午 4:30 到乒乓球大厅受训，武汉的乒乓球健将们将为你们做专门指导。

校体育办公室<br>2009年9月10日

---

Dear Mr. Harrison,

Our new factory will be commencing production on April 10 and we should like to invite you and your wife to be present at a celebration to mark the occasion.

As you will appreciate this is an important milestone for this organization, and is the result of continued demand for our products, both at home and overseas. We are inviting all those individuals who contribute to the company's success and trust that you will pay us the compliments of accepting.

Please confirm that you will be able to attend by advising us of your time — we can arrange for you to be met. All arrangements for your stay will, of course, be made by us at our expense.

Yours faithfully,

---

✗　　亲爱的哈里森先生，

　　本公司新厂将于4月10日开始投产，希望能邀请您和您的太太来参加新厂开工典礼。

　　如您所知，新厂的设立是本公司的一个里程碑，而这正是海内外客户对本公司产品不断需求的结果。我们邀请了所有对本公司的成功贡献力量的人，我们相信，您一定会赏光。

　　如您确能参加，请来函告知您抵达的时间，以便我们为您安排会晤。当然，在此期间所有安排及费用皆将由本公司承担。

您诚挚的

# Practice

**NOTICE**

_____

_____

_____

_____

ABC Managing Director: Jack. Smith

✉  通　　知

我们将于5月10日在本公司会议室举行会议，讨论今年下半年的销售计划，请各部门的负责人参加。如果不能参加，请务必事先通知我们。

ABC 公司总经理：杰克·史密斯

Dear Mr. Smith,

_____

_____

_____

_____

Sincerely yours,

✉  尊敬的史密斯先生：

如您能够出席为中国代表团而举行的招待会，我们将感到十分荣幸。

招待会定于 10 月 4 日(星期二)在市政厅举行。下午 6 点钟准时举行鸡尾酒会，随之在 8 点钟举行晚宴。

我们期待着您的光临。请提前告知我们您能否出席。

您诚挚的

# Tips

### 几种常见的邀请信

☒ Dear sir/madam,

I'm delighted that you have accepted our invitation to speak at the Conference in [city] on [date].

As we agreed, you'll be speaking on the topic... from [time] to [time]. There will be an additional minutes for questions.

Would you please tell me what kind of audio-visual equipment you'll need? If you could let me know your specific requirements by [date], I'll have plenty of time to make sure that the hotel provides you with what you need.

Thank you again for agreeing to speak. I look forward to hearing from you.

Sincerely yours,

[Name]

[Title]

☒ Dear sir/madam,

Thank you for your letter of [date]. I'm glad that you are also going to [place] next month. It would be a great pleasure to meet you at the [exhibition/trade fair].

Our company is having a reception at [hotel] on the evening of [date] and I would be very pleased if you could attend.

I look forward to hearing from you soon.

Yours sincerely,

[Name]

[Title]

✉ Dear sir/madam,

[Organization] would very much like to have someone from your company speak at our conference on [topic].

As you may be aware, the mission of our association is to promote. Many of our members are interested in the achievements your company has made in.

Enclosed is our preliminary schedule for the conference which will be reviewed in weeks. I'll call you [date] to see who from your company would be willing to speak to us. I can assure you that well make everything convenient to the speaker.

Sincerely yours,

[Name]

[Title]


✉ Dear sir/madam,

We would like to invite you to an exclusive presentation of our new [product]. The presentation will take place at [location], at [time] on [date]. There will also be a reception at [time]. We hope you and your colleagues will be able to attend.

[Company] is a leading producer of high-quality. As you well know, recent technological advances have made increasingly affordable to the public. Our new models offer superb quality and sophistication with economy, and their new features give them distinct advantages over similar products from other manufacturers.

We look forward to seeing you on [date]. Just call our office at [phone number] and we will be glad to secure a place for you.

Sincerely yours,

[Name]

[Title]


✉ Dear sir/madam,

On [date], we will host an evening of celebration in honor of the retirement of [name], President of [company]. You are cordially invited to attend the celebration at [hotel], [location], on [date] from to p.m.

[name] has been the President of [company] since [year]. During this period, [company] expanded its business. Now it's our opportunity to thank him for his years of exemplary leadership and wish him well for a happy retirement. Please join us to say Good-bye to [name].

See you on [date].

Yours sincerely

[Name]

[Title]

# UNIT 10

# TOPICS

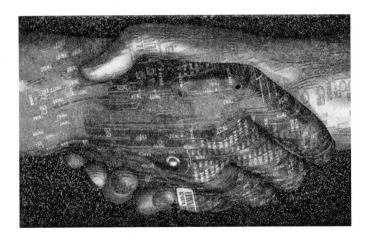

- What is embedded system?
- What is the history of embedded system?
- How about the difference between common PC and embedded system?
- What does embedded system do?
- What are the characteristics of embedded system?
- What do you think of the future of embedded system?
- Understanding party culture in the West.
- How to write the abstracts of scientific and technological theses?

 TEXT

# Embedded System

An embedded system is a computer system designed for **specific** control functions within a larger system, often with real-time computing **constraints**. It is embedded as part of a complete device often including hardware and mechanical parts. By contrast, a general-purpose computer, such as a personal computer (PC), is designed to be **flexible** and to meet a wide range of end-user needs. Embedded systems control many devices in common use today.

Embedded systems contain processing cores that are typically either **microcontrollers** or digital signal processors (DSP). The key characteristic, however, is being **dedicated** to handle a particular task. Since the embedded system is dedicated to specific tasks, design engineers can **optimize** it to reduce the size and cost of the product and increase the reliability and performance. Some embedded systems are **mass-produced**, benefiting from economies of scale.

Physically, embedded systems range from portable devices such as digital watches and MP3 players, to large **stationary** installations like traffic lights, factory controllers, or the systems controlling nuclear power plants. Complexity varies from low, with a single microcontroller chip, to very high with multiple units, **peripherals** and networks mounted inside a large **chassis** or **enclosure**.

## History

Small systems but still required many external memory

specific [spi'sifik] adj. 特殊的，特定的

constraint [kən'streint] n. 约束；限制

flexible ['fleksəbl] adj. 可弯曲的，易弯曲的；柔韧的；有弹性的

microcontroller ['maikrəukən'trəulə] n. 微型控制器
dedicated ['dedi'keitid] adj. (电脑)专用的
optimize ['ɔptimaiz] vi. 使最优化

mass-produced 大(批)量生产的

stationary ['steiʃənəri] adj. 不动的；不增减的

peripheral [pə'rifərəl] adj. (电脑的)外部设备
chassis ['ʃæsi] n. (装置的)底架(盘)
enclosure [in'kləuʒə] n. 外壳，套

and support chips. In 1978 National Engineering Manufacturers Association released a "standard" for programmable microcontrollers, including almost any computer-based controllers, such as single board computers, numerical, and event-based controllers.

As the cost of microprocessors and microcontrollers fell it became feasible to replace expensive analog components such as **potentiometers** and variable capacitors with up/down buttons or **knob**s read out by a microprocessor even in some consumer products. By the mid-1980s, most of the common previously external system components had been integrated into the same chip as the processor and this modern form of the microcontroller allowed an even more widespread use, which by the end of the decade were the **norm** rather than the exception for almost all electronics devices.

The integration of microcontrollers has further increased the applications for which embedded systems are used into areas where traditionally a computer would not have been considered. A general purpose and comparatively low-cost microcontroller may often be programmed to fulfill the same role as a large number of separate components. Although in this context an embedded system is usually more complex than a traditional solution, most of the complexity is contained within the microcontroller itself. Very few additional components may be needed and most of the design effort is in the software. The **intangible** nature of software makes it much easier to prototype and test new revisions compared with the design and construction of a new circuit not using an embedded processor.

## Characteristics

1. Embedded systems are designed to do some specific task, rather than be a general-purpose computer for multiple tasks. Some also have real-time performance constraints that must be met, for reasons such as safety and usability; others may have low or no performance

requirements, allowing the system hardware to be simplified to reduce costs.

2. Embedded systems are not always standalone devices. Many embedded systems consist of small, computerized parts within a larger device that serves a more general purpose. For example, the Gibson Robot Guitar **features** an embedded system for tuning the strings, but the overall purpose of the Robot Guitar is, of course, to play music. Similarly, an embedded system in an automobile provides a specific function as a subsystem of the car itself.

feature [fi:tʃə] vi. 以……为特色；是……的特色

3. The program instructions written for embedded systems are referred to as firmware, and are stored in read-only memory or Flash memory chips. They run with limited computer hardware resources: little memory, small or **non-existent** keyboard or screen.

non-existent 不存在的

# EXERCISES

Ⅰ. Match the terms and the interpretations.

1. Microcontroller Unit (MCU)   (a) A category of software tools for designing electronic systems such as printed circuit boards and integrated circuits. The tools work together in a design flow that chip designers use to design and analyze entire semiconductor chips.

2. Reduced Instruction Set Computing (RISC)   (b) Apple Inc.'s mobile operating system. Originally developed for the iPhone and iPod Touch, it has since been extended to support other Apple devices such as the iPad, and Apple TV.

3. Android   (c) A single chip that contains a processor, RAM, ROM, clock and I/O control unit. Hundreds of millions of MCUs are used in myriad devices ranging from automobiles to action figures.

4. iOS

(d) A CPU design strategy based on the insight that simplified (as opposed to complex) instructions can provide higher performance if this simplicity enables much faster execution of each instruction. A computer based on this strategy is a reduced instruction set computer also called RISC.

5. Electronic Design Automation (EDA)

(e) A Linux-based operating system for mobile devices such as smartphones and tablet computers. It is developed by the Open Handset Alliance led by Google.

II. Are the following statements True (T) or False (F)?

1. (    ) An embedded system is a general-purpose system.

2. (    ) Physically, embedded systems range from portable devices such as digital watches and MP3 players, to large stationary installations like traffic lights, factory controllers.

3. (    ) Embedded systems range from no user interface at all — dedicated only to one task — to complex graphical user interfaces that resemble modern computer desktop operating systems.

4. (    ) Apple licenses iOS for installation on non-apple hardware.

5. (    ) Embedded software's principal role is not information technology, but rather the interaction with the physical world.

6. (    ) Embedded software is usually written for special purpose hardware.

7. (    ) "Microcontroller" is a synonym of "microprocessor".

8. (    ) The relative simplicity of ARM processors makes them suitable for low power applications.

9. (    ) Machines that embedded systems reside in make no errors at all.

10. (    ) Unreliable mechanical moving parts such as disk drives, switches or buttons are often used in embedded system

III. Translate the following words and phrases into Chinese.

1. Embedded Technology            _____

2. Embedded Real-Time OS          _____

3. Boot Loader　　　　　　　　　＿＿＿＿＿＿＿＿＿＿
4. Microcontroller　　　　　　　　＿＿＿＿＿＿＿＿＿＿
5. I/O Space　　　　　　　　　　＿＿＿＿＿＿＿＿＿＿
6. Interrupt Vector　　　　　　　　＿＿＿＿＿＿＿＿＿＿
7. Kernel　　　　　　　　　　　　＿＿＿＿＿＿＿＿＿＿
8. Multitasking　　　　　　　　　＿＿＿＿＿＿＿＿＿＿
9. DSP　　　　　　　　　　　　　＿＿＿＿＿＿＿＿＿＿
10. System Response Time　　　　　＿＿＿＿＿＿＿＿＿＿

Ⅳ. Translate the following Chinese statements into English.

1. 嵌入式系统把计算机直接嵌入到应用系统中，它融合了计算机软/硬件技术、通信技术和微电子技术。

2. 20世纪70年代单片机的出现，使得汽车、家电、工业机器、通信装置以及成千上万种产品可以通过内嵌电子装置来获得更佳的使用性能。

3. 嵌入式系统的构架可以分成四个部分：处理器、存储器、输入/输出(I/O)单元和软件。

4. 嵌入式实时系统以其简洁、高效等特点越来越多地受到人们的广泛关注。

5. 嵌入式操作系统相关的研究包括嵌入式操作系统的实时性、通用性、扩展性和安全性。

Ⅴ. Fill in each of the blanks with one of the following words or phrases.

*storage　independently　real-time　software　different　dedicated*

*compose　interaction*

Embedded System refers to a kind of device or appliance that only has the functions of a computer, but not a PC system. It centers on application and its software & hardware can be downsized so as to cater to the_____computer systems that have strict requirements on the function, reliability, cost, volume and power consumption of the application system. Simply speaking, with its applications and hardware all in one, the embedded system resemble the way that BIOS functions in a PC, its high-automation and quick-response makes it especially suitable for a _____and multi-task computer system. This system is_____of embedded processor, supporting hardware, embedded O/S and its application_____. It is a kind of device that can function_____. The hardware part consists of processor or microprocessor, _____, peripheral devices, I/O ports and graphic controller, etc. Embedded system is_____from a general computer processing system, as it doesn't have such a large storage medium as HDD in a PC, only supporting EPROM/EEPROM or Flash Memory. The software consists of operating system (request for real-time & multi-task operation) and the application programming. The applications control the operating and functioning of the system, while the operating system controls the_____between the application programming and the hardware.

# SUPPLEMENTARY

## Inception Review

"True inspiration is impossible to fake", explains a character in Christopher Nolan's existentialist heist film Inception. If that's the case, then Inception is one of the most honest  films ever made. Nolan has crafted a movie that's beyond brilliant and layered both normatively and thematically. It requires the audience to take in a collection of rules, exceptions, locations, jobs, and abilities in order to understand the text, let alone the fascinating subtext. Nolan's magnum opus is the first major blockbuster in over a decade that's demanded intense viewer concentration, raised thoughtful and complex ideas, and wrapped everything all in a breathlessly exciting action film. Inception may be complicated, but simply put it's one of the best movies of the year.

Inception requires so much exposition that a lesser director would have forced theaters to distribute pamphlets to audience members in order to explain the complicated world he's developed. During my first draft of this view, I realized I had spent three paragraphs simply trying to explain the plot. I will simply avoid this exposition and present the movie's basic premise. Inception centers on a team of individuals led by an "extractor" named Cobb (Leonardo DiCaprio) who, through the use of a special device, construct the dreams of a target and use those dreams to implant an idea so that the target will make a decision beneficial to the individual who hired the team. To say that scratches the surface would be an insult to both scratches and surfaces. But since it takes Nolan about fifty minutes to set everything up, I hope you'll forgive my brevity.

Why is it so difficult to explain the plot in depth? First, I don't want to spoil you. Secondly, the film layers dreams on top of dreams to the point where a unique keepsake called a "totem" is required in order to inform a character as to whether or not he or she is still dreaming. Then you have people in particular roles like "The Architect", "The Forger", and "The Chemist" in order to pull off the job. Furthermore, dreams have rules: dying in a dream forces the dreamer to wake up, delving too deeply into a mind can cause an eternal slumber called

"Limbo", using memories to construct dreams is dangerous because it can blur the line between dreams and reality. In addition, intruding in the dreams of another will cause the dreamer's "projections" (human representations created by the dreamer) to attack the intruders like white blood cells going after an infection. And these explanations only represent a fraction of the terminology, rules, exceptions, or details that are necessary for creating the world of Inception.

But it's not a confusing movie if you provide it with your full attention. There are a lot of summer movies that ask you turn off your brain and enjoy the persistent-vegetative-state ride. Inception is not one of those movies. There's a lot to take in, but the imaginative and thoughtful delivery of exposition keeps the viewer riveted despite the amount of information required in order to understand the premise, setting, and plot.

As you've probably guessed, when I said at the beginning of this review that Inception was the first movie in over a decade to mix breathtaking action with thoughtful subtext, I was referring to 1999's The Matrix. The comparisons are inevitable. Both movies deal with the nature of reality combined with pulse-pounding set pieces that will be included in any action-scene highlight reel. But The Matrix is a freshman level course compared to the doctorate held by Inception, and it has nothing to do with how far special effects have come in ten years. It's about taking multiple genres, settings, ideas, emotions, and questions and weaving them into a rich tapestry that will have folks talking long after the credits roll. But then you throw in those advanced special effects and you have a summer blockbuster that will blow your mind. You've never seen anything like Inception, and you'll want to see it again and again.

# CONVERSATION

## Understanding Party Culture in the West

## 了解西方的 Party 文化

Party 在这里不指"党派",而是指各种聚会。许多西方人热衷于举办各种 Party,一有适当的机会就会聚在一起乐一下。除了大家比较熟悉的生日聚会(Birthday Party)、结婚宴会(Wedding Party)、舞会(Ball Party)等,还有许多其他形式、种类的聚会,现将其中一些聚会及其习惯、习俗介绍给大家。

1. 赏秋聚会(Autumn Foliage Party)

秋天,美国东北部漫山遍野的树林颜色变幻无穷,朋友们经常相邀欣赏迷人的秋色,故得此名。

2. 棕色纸袋会(Brown Bag Party)

这是一种自带食品的聚会。在这种聚会上(一般是午餐时间),大家各自吃自己棕色纸袋里带去的食品,轻松自如地随意交谈。因为美国食品店都是用牛皮纸包装食品,是一种经济实惠的交际活动,故得此名。

### 3. 樱桃聚会(Cherry Party 或 Cherry Hour)

该聚会实质上是工作交流聚会,一般在下午三四点钟,一天工作即将结束时举行。各部门有关人员聚在一起,边喝饮料边交流当天的工作情况。

### 4. 圣诞树装饰会(Christmas Tree Decorating Party)

这是为迎接圣诞节各家各户举办的活动,边唱边跳边装饰,节日气氛浓厚。参加者需带绸带、剪纸、纸花等装饰品。

### 5. 鸡尾酒会(Cocktail Party)

这种聚会常在下午三至四时,或五至七时举行。客人可以晚到或早来,来去比较自由。在这种聚会上人们品尝各种鸡尾酒,故得此名。

### 6. 欢送会(Farewell Party)

欢送会是在办公室里占用上班时间开的,一般是买点礼物点些 Pizza 即可。

### 7. 暖屋会(Housewarming Party)

"暖屋"从字面上解释就是"把屋子弄热",暖屋会一般是搬了新家以后开,给新家增加人气。与会者常带上一两件小礼物,如炊具等家庭实用物品,以帮助主人开始新的生活。

### 8. 费用分担的社交聚会(No-host Party)

no-host 是"没有主人"的意思。这种社交聚会由参加的人各自分担费用,形式上有主办者,但没有担负全部费用的主人(host)。大家可分担现金,也可各自带食品和饮料。

### 9. 野炊聚会(Picnic Party 或 Cook-Outs)

野炊是美国人最喜欢的活动之一。全家人或几家人带上炊具、食物到野外边烧烤边吃边聊天,别有一番情趣。

### 10. 家常聚餐会(Potluck Party)

这是美国人最经常举行的一种典型聚会,源于早年农忙季节,邻居互相帮工,并把从自家带来的食物放在一起共同享用。现在这种聚会则是个人或各家带着足够几个人或几家人吃的食品(通常是自己做的拿手好菜)聚到一起,分享各家的精美食品。

### 11. 大家缝聚会(Quilting Party 或 Quilting Bee 或 Quilting)

早些时候,美国妇女很少外出工作,无聊时就带上要缝的被子聚在一起边缝边闲谈,故得此名。

### 12. 男士聚会(Stag Party)

Stag Party 只限于男子参加,女性一律谢绝。在 Stag Party 上,男士们主要是打扑克,一起看拳击赛或者足球赛。有时候,会在新郎举行婚礼的前夜举行 Stag Party,以纪念他告别单身生活。

### 13. 超级杯聚会(SuperBowl Party)

这是在超级杯总决赛的时候开的聚会，一般不管是不是球迷都会参加，因为一年里最精彩的新广告会在比赛中间播出。

### 14. 惊喜聚会(Surprise Party)

聚会主角一般事先不知情，被骗到现场后大家从藏身之处跳出来大喊："SURPRISE!"。大多是为某人生日、搬家、升迁等举行。

### 15. 车尾野餐会(Tailgate Party)

这是一种不分男女老少的聚会。tailgate 原是一种箱形轿车尾部的车门，它可以翻下来当桌子用。一些球迷在比赛开始前几小时就捷足先登，把三明治、热狗或者烘烤肉类放在这张车尾的临时桌子上举行野餐，这就是 Tailgate Party 的由来。

# Example

## ✉ Dialogue 1

A: Are you going to Helen's birthday party on Friday evening?

B: I wouldn't miss it for the world! It's sure to be fun. She's invited a lot of people. Do you think everyone will be able to get into her house?

A: If everyone turned up，it would be a squeeze，but a few people said that they couldn't go, so I think it should be OK.

B: Are you taking anything?

A: I've got her a birthday present and I'll take a bottle of wine too.

B: That's a good idea. She told she had bought plenty of food and snacks. I think it's going to be a noisy party. I hope her neighbors don't mind too much.

A: Helen gets on very well with her neighbours. I wouldn't be surprised if they went to the party too.

B: I'm really looking forward to it. This party is going to be a blast!

A: Well，don't be late. I'll see you on Friday at Helen's.

## ✉ Dialogue 2

A: Can I get you something to drink?
B: No, that's OK, I already have a coke. Why don't you have a seat? You look like you've been on your feet all day.
A: I guess I could take a break. So, how do you like Denver?
B: It's great! The mountains here are beautiful, and the skiing is spectacular. Have you been here for a long time?
A: About six years...
B: What do you do for a living?
A: I manage one of the ski lodges. It's a great job, I can spend a lot of my time outdoors, and I also get to ski for free all season.
B: Wow, talk about job perks! That's great. I would like to do something exciting like that. But I am only an accountant. Not too much excitement there, huh?
A: That's okay. If it weren't for you accountants, nobody would have the money to go skiing!

## Practice

Imagine that you are at a company party where you are supposed to talk with a new colleague. Practice the following conversation with the given points:

- ✧ Introduce your roles in the company.
- ✧ Learn about your hometowns.
- ✧ Talk about your hobbies.
- ✧ Exchange contact information with each other.

## Tips

## ✉ Party 常用语

1. I am throwing a party Saturday. Would you like to come? 周六我要办个聚会，你来吗？

2. I'd love to, but I already have plans. 我很想去，但是我已经有安排了。

3. It's a surprise party. 这是一个惊喜聚会。

4. Are you alone? 你是一个人来吗？

5. How many people are coming? 共有多少人要来呢?

6. Who is going to organize the birthday party next time? 下次轮到谁办生日聚会了呢?

7. I am turning 23 tomorrow. 明天我就二十三岁了。

8. Are you guys OK? 你们都还好吧?

9. Can I bring anything to the party? 我要带点什么去参加聚会呢?

10. Make yourself at home! 请别客气.

11. Everybody picks up whatever you want. 每个人拿任何你们想要的东西。

12. You are a party pooper. 你真是扫兴。

13. He dances like an animal. 他跳舞跳得很疯狂。

14. That guy is such a party animal. 那家伙真是个聚会狂。

15. I really enjoyed the food and music tonight. 我真的很喜欢今晚的食物和音乐。

16. This place is so cool! 这个地方真不错!

17. Do you have a good time today? 今天玩得高兴吗?

18. I'll walk you out. 我送你出去。

19. This party is so dull. I don't know anyone here, who are these people? 这个聚会太无聊了,我谁都不认识,这些人都是谁啊?

20. That's a real eye-opening experience. 那真是一个令人大开眼界的经历。

 WRITING

# How to Write the Abstracts of Scientific and Technological Theses

## 如何写科技论文摘要

为适应世界范围内科技文献的快速增加,方便科技工作者检索,科技文献一般要附有英文摘要。摘要又称提要、文摘,其对应的英文名称为 abstract。国际标准 ISO214-76 对其的解释为：an abbreviated, accurate representation of the contents of a document without added interpretation or criticism. 意思是对文献内容的准确压缩，不加以解释或评论。

一般所说的英文摘要是指"英文摘要正文"。而将英文标题、作者姓名汉语拼音、其工作单位英译文、英文摘要正文、英文关键词放在一起统称为"英文摘要"。绝大多数有英文摘要的文章一般把这五个部分放在一起作为相对独立于文章正文的一个单独的部分。

### 1. 摘要的种类

科技论文英文摘要主要有两种类型：

(1) 信息型摘要(Informative Abstract)：也称报道型摘要，主要用于实验性和技术性较强的论文，报道论文的研究成果、数据和结论，对于最佳条件、成功的数据及误差范围、结论及适用范围如实给出。其中要综述论文的主要内容、要旨、重点，还需列出有关的具体数据、实验结果以及采用的方法。信息型摘要多用于科技杂志或科技期刊的文章，也用于会议论文及各种专题技术报告。这类摘要可分段。

(2) 指示性摘要(Indicated Abstracts)：也称描述性摘要，多用于理论性较强的论文，如专论、评论、综述、数学计算、理论推导等，主要概括论文的主要论点、分析过程和结论。该类摘要很少传递具体数据，而只告诉读者本文采用了什么方法、讨论了什么问题、得出了什么结论，内容宏观，篇幅较短。综述性、评论性或资料性论文宜写成指示性摘要。这类摘要一般不分段。

### 2. 摘要的四要素

(1) 目的(Objective)：概括研究的前提、目的、任务及所涉及的主题范围，该项内容通常是用一句话来完成，必须紧扣论文题目和主题。通常采用一般现在时态或直接用动词不定式词组来表达。

(2) 方法(Methods)：陈述研究实验中的对象，实验研究中曾使用的材料、设备、工艺、手段、程序等方面。该项内容包括较多，组织较难，一定要用较为简短的语言来表达其内容，而初学者往往不能控制且浓缩不好，总希望将实验或研究的全部内容都加以表达，导致篇幅较大，最终事与愿违，效果不佳。因该项目所表述的是已经采用的实验方法、路线和手段，故常用一般过去时态。

(3) 结果(Results)：揭示已经得出的研究结果，包括数据、效果、性能等方面。该项内容用一句话来概括通过方法中的手段得到的结果以及研究目的的达到与否，同时对研究者已经完成的研究进行总结，因此常用一般过去时态。

(4) 结论(Conclusion)：根据研究结果提出问题、建议、预测，包括对结果的分析、比较、应用等方面。此项内容包括通过实验研究或总结，对于同行和阅读者有什么帮助、建议和启迪，因此常用一般现在时态或情态动词。

### 3. 摘要的文体特点

摘要的写作原则是提纲挈领，重点突出，内容完整，语言准确、客观和简洁。语体正式、句法结构规范、用词准确精炼、篇章结构紧凑完整是科技论文英文摘要的文体特点。写作科技论文英文摘要时，应多用规范正式的专业词汇或使用范围较窄、意义更准确的"大词",多用现在时态和被动语态。

(1) 正规。摘要一般以专业人员为读者对象,属于正式文体,句法结构要求严谨规范。因此,摘要中的句子都很完整,没有口语体中的省略句或不完整句。用词也很规范,多用论文研究领域的标准术语、正规英语,很少用缩写词和古词。

(2) 精炼。摘要要求精炼,不宜列举例证,不宜与其他研究工作作对比,语句也少有重复。在衔接方面,主要使用词汇手段,通过词汇在意义上的衔接,把全篇文章的各部分紧紧地联系在一起,使文章结构紧凑,前后呼应。复合名词可以使文字紧凑利落,因而摘要中复合名词用得较多。

(3) 具体。摘要的每个概念、论点都要具体鲜明。一般不笼统地写论文"与什么有关",而直接写论文"说明什么"。用词方面要求准确,多用一些源自法语和拉丁语且使用范畴较窄的"大词"、"长词",尽量避免含混不清或一词多义的词语。

(4) 完整。摘要本身要完整。有些读者是利用摘要杂志或索引卡片进行研究工作的,很可能得不到全篇论文,因此要注意不要引用论文某节或某张插图来代替说明。

# Example

## Applying Matlab Stochastic Models on Computer Application Systems via CORBA

Quan Xiaohong

(Software College, Changzhou College of Information Techology, Changzhou, Jiangsu, 213164, China)

**Abstract:** A framework of applying Matlab stochastic models on computer application systems via CORBA interface was proposed. The results of implementing an option pricing model on an energy trading system indicate that this generic solution is able to improve the system executing efficiency.

**Key words:** Matlab; CORBA; stochastic model; energy trading system

## 基于 CORBA 的 Matlab 随机模型在计算机应用系统中的应用

权小红

(常州信息职业技术学院软件学院,中国江苏 常州 213164)

**摘要:** 介绍一种基于 CORBA 接口的 Matlab 随机模型在计算机应用系统上的实现方案。应用该方案在能源交易系统上进行的期权定价模型实验结果表明，该方案能有效地缩短系统运行时间，具有一定的通用性。

**关键词**：Matlab；CORBA；随机模型；能源交易系统

## Edification of the Development of Indian Software Industry on Chinese Higher Vocational Software Talent Cultivation

Quan Xiaohong, Tang Xiaoyan

(Software College, Changzhou College of Information Techology, Changzhou, Jiangsu, 213164, China)

**Abstract**: The rapid development and success of Indian software industry are of vital importance to our country. Based on the actuality of higher vocational school in China, we should change our ideas of teaching, set the proper training objective of higher vocational software talents, explore scientifically, develop the courses properly, focuses on practical work, follow the teaching mode of "learning by doing"; comprehensively improve the qualifications of software talents to meet with the severe international competition.

**Key words**: Indian software industry; higher vocational software talents; cultivation mode

## 印度软件业发展对我国高职软件人才培养的启示

权小红，唐小燕

(常州信息职业技术学院软件学院，中国江苏 常州 213164)

**摘要**：印度软件业的迅猛发展与成功经验，对我国高职软件人才培养有重要的借鉴意义。结合目前我国高职软件学院实际，应转变观念，树立高职软件人才培养目标，科学论证，合理开发课程，注重实践，开展"做"中"学"的教学模式，全面提高软件人才职业素质，以适应日益激烈的国际竞争需要。

**关键词**：印度软件业；高职软件人才；培养模式

## Practice

### Application of Automatically Booking Air Tickets on ATM

Quan Xiaohong

(Software College, Changzhou College of Information Techology, Changzhou, Jiangsu, 213164, China)

**Abstract:** Defines _____
_____ , and illustrates the design and application concept of this system including the designing of the system architecture, the workflow of listening in monitor, and the solutions for_____ .Also concludes _____.

**Key words:** ATM; _____; bank card payment;_____

### ATM 民航自助售票系统的设计与实现

权小红

(常州信息职业技术学院软件学院，中国江苏 常州 213164)

**摘要**：明确了 ATM 民航自助售票系统的系统功能需求，提出了系统设计实施方案，主要包括系统结构设计、守候服务主要工作流程以及关于系统并发性、安全性等技术要点的解决策略。最后总结了系统优点及实用价值。

**关键词**：ATM；自动售票系统；银行卡支付；守候服务

### Study on the Comprehensive Evaluation of Teaching Effectiveness in High Vocational Education

Quan Xiaohong, Li Chunhua

(Software College, Changzhou College of Information Techology, Changzhou, Jiangsu, 213164, China)

**Abstract:** How to improving faculties' teaching effectiveness in high education so as to _____has been a important issue in the development of high education._____
_____
_____.

**Key words:** _____; teaching effectiveness; comprehensive evaluation

## 高职教师教学质量综合评价研究

权小红 李春华

(常州信息职业技术学院软件学院，中国江苏 常州 213164)

**摘 要**：如何提高高职院校教学质量，加强对教师教学质量的科学评价与管理，已成为高职院校发展中面临的重要课题。应充分认识高职教师教学质量综合评价的含义与功能，明确综合评价的主体与评价指标的设置原则，确立综合评价的科学方法，最终构建高职院校教师教学质量综合评价体系。

**关键字**：高职院校；教学质量；综合评价

## Tips

### 英文摘要常用句型

1. This paper gives a brief introduction to ...

2. This paper discusses ...

3. This article explores...

4. This paper provides a method of ...

5. This paper introduces an applicable procedure to analyze...

6. This paper offers the latest information regarding...

7. This article gives a complete commentary on the ...

8. In this paper are presented the results ...

9. A new method is described for ...

10. The principle of ... is outlined.

11. The use of ... is addressed.

12. The mechanism of ... is examined.

13. The dependence of ... was established.

14. An analysis of ... was carried out.

15. The aim of this study is to ...

16. The problem of something is discussed...

17. The study of ... is based on ...

18. The idea of our method of measurement is to analyze ...

19. Comparison of our results with ...

20. We thus conclude that ...

21. The result shows that ...

22. This article shows that...

23. It is suggested that ...

24. The author's suggestion (or conclusion) is that ...

25. Although a number of tests and comparison of the method have given satisfactory results, additional investigations to provide further justification and verification are required.

# 附录 计算机英语常用词汇术语表

## A

Access Control List(ACL) 访问控制列表

access token 访问令牌

access 访问

account lockout 账号封锁

account policies 记账策略

active file 活动文件

active 激活

adapter 适配器

adaptive speed leveling 自适应速率等级调整

Address Resolution Protocol(ARP) 地址解析协议

add watch 添加监视点

administrator account 管理员账号

algorithm 算法

allocation layer 应用层

allocation 分配、定位

all rights reserved 所有的权力保留

ANSI(American National Standards Institute) 美国国家标准协会

API (Application Program Interface) 应用程序编程接口

archive file attribute 归档文件属性

ARPANET 阿帕网(Internet 的前身)

ASP(Active Server Pages) 活动服务器页面(一个编程环境,在其中,可以混合使用 HTML、脚本语言以及组件来创建服务器端功能强大的 Internet 应用程序)

assign to 指定到

ATM (Asynchronous Transfer Mode) 异步传输模式

attack 攻击

attribute 属性

auditing 审计、监察

authentication 认证、鉴别

authorization 授权

auto answer 自动应答

auto detect 自动检测

auto indent 自动缩进

auto save 自动存储

## B

back up 备份

back-end 后端

backup browser 后备浏览器

bad command 命令错

bad command or file name 命令或文件名错

baseline 基线

batch parameters 批处理参数

BDC(Backup Domain Controller) 备份域控制器

BGP(Border Gateway Protocol) 边界网关协议

binary file 二进制文件

binding 联编、汇集

BIOS(Basic Input/Output System) 基本输入/输出系统

BOOTP 引导协议

border gateway 边界网关

bottleneck 瓶颈

bottom margin 页下空白

breach 攻破、违反

breakable 可破密的

bridge 网桥、桥接器
browser 浏览器
browsing 浏览
by extension 按扩展名
bytes free 字节空闲

# C

class A domain A 类域
class B domain B 类域
class C domain C 类域
call stack 调用栈
case sensitive 区分大小写
CD-ROM 光盘驱动器(光驱)
CD-R (Compact Disk-Recordable) 可擦写光盘
CGI (Common Gateway Interface) 公共网关接口
   (Computer Graphics Interface) 计算机图形接口
CGI-based attack 基于CGI攻击(它利用公共网关接口的脆弱点进行攻击，通常借助www站点进行)
change directory 更换目录
change drive 改变驱动器
change name 更改名称
channel 信道、通路
character set 字符集
checks a disk and displays a status report 检查磁盘并显示一个状态报告
checksum 校验和
chip 芯片
choose one of the following 从下列中选一项
cipher 密码
cipher text 密文
CIX(Commercial Internet Exchange) 两个商业ISP的连接点

classless addressing 无类地址分配

clear all breakpoints 清除所有断点

clears an attribute 清除属性

clears command history 清除命令历史

clear screen 清除屏幕

clear text 明文

client 客户，客户机

client/server 客户机/服务器

close all 关闭所有文件

cluster 簇、群集

CMOS(Complementary Metal-Oxide-Semiconductor) 互补金属氧化物半导体

code generation 代码生成

color palette 彩色调色板

column 行

COM port COM 口(通信端口)

command line 命令行

command prompt 命令提示符

component 组件

compressed file 压缩文件

computer language 计算机语言

configuration 配置

configures a hard disk for use with MS-DOS 配置硬盘，以为 MS-DOS 所用

conventional memory 常规内存

copies files with the archive attribute set 拷贝设置了归档属性的文件

copies one or more files to another location 把文件拷贝或搬移至另一地方

copies the contents of one floppy disk to another 把一个软盘的内容拷贝到另一个软盘上

copy diskette 复制磁盘

copyright 版权

CPU(Center Processor Unit) 中央处理单元

crack 闯入

crash 崩溃(系统突然失效，需要重新引导)

create DOS partition or logical DOS drive 创建 DOS 分区或逻辑 DOS 驱动器

create extended DOS partition 创建扩展 DOS 分区

create logical DOS drives in the extended DOS partition 在扩展 DOS 分区中创建逻辑 DOS 驱动器

create primary DOS partition 创建 DOS 主分区

creates a directory 创建一个目录

creates changes or deletes the volume label of a disk 创建、改变或删除磁盘的卷标

cruise 漫游

cryptanalysis 密码分析

CSU/DSU(Channel Service Unit/Data Service Unit) 通道服务单元/数据服务单元

current file 当前文件

current fixed disk drive 当前硬盘驱动器

current settings 当前设置

cursor position 光标位置

cursor 光标

## D

default route 缺省路由

default share 缺省共享

database 数据库

data link 数据链路

data-driven attack 数据驱动攻击(依靠隐藏或者封装数据进行的攻击，那些数据可不被察觉的通过防火墙)

datagram 数据报

DBMS(DataBase Management System) 数据库管理系统(一种操纵和管理数据库的大型软件，用于建立、使用和维护数据库)

DDE(Dynamic Data Exchange) 动态数据交换

debug 调试

decryption 解密

default document 缺省文档

default 默认

defrag 整理碎片

delete partition or logical DOS drive 删除分区或逻辑 DOS 驱动器

deletes a directory and all the subdirectories and files in it 删除一个目录和所有的子目录及其中的所有文件

demo 演示

denial of service 拒绝服务

destination folder 目的文件夹

device driver 设备驱动程序

DHCP(Dynamic Host Configuration Protocol) 动态主机配置协议

dialog box 对话框

dictionary attack 字典式攻击

digital key system 数字键控系统

direction keys 方向键

directory replication 目录复制

directory list argument 目录显示变量

directory structure 目录结构

disk mirroring 磁盘镜像

disk access 磁盘存取

disk copy 磁盘拷贝

disk space 磁盘空间

display options 显示选项

display partition information 显示分区信息

displays files in specified directory and all subdirectories 显示指定目录和所有目录下的文件

displays files with specified attributes 显示指定属性的文件

displays or changes file attributes 显示或改变文件属性

displays or sets the date 显示或设置日期

displays setup screens in monochrome instead of color 以单色而非彩色显示安装屏信息

displays the amount of used and free memory in your system  显示系统中已用和未用的内存数量

displays the full path and name of every file on the disk  显示磁盘上所有文件的完整路径和名称

displays the name of or changes the current directory  显示或改变当前目录

distributed file system  分布式文件系统

DLC(Data Link Control)  数据链路控制

DNS(Domain Name System)  域名系统(在 Internet 上查询域名或 IP 地址的目录服务系统)

DNS spoofing  域名服务器电子欺骗(攻击者用来损害域名服务器的方法，可通过欺骗 DNS 的高速缓存或者内应攻击来实现的一种方式)

domain controller  域名控制器

domain name  域名

DOS shell  DOS 外壳

## E

eavesdropping  窃听、窃取

edit menu  编辑选单

EGP(Exterior Gateway Protocol)  外部网关协议

E-mail  电子邮件

EMS(Expanded Memory System)  扩充内存

encrypted tunnel  加密通道

encryption  加密

end of file  文件尾

end of line  行尾

enter choice  输入选择

enterprise network  企业网

entire disk  转换磁盘

environment variable  环境变量

Ethernet  以太网

every file and subdirectory  所有的文件和子目录

exception  异常

execute 执行

expand tabs 扩充标签

explicitly 明确地

extended memory 扩展内存

external security 外部安全性

# F

FAT(File Allocation Table) 文件分配表

fax modem 传真猫

FDDI (Fiber Distributed Data Interface) 光纤分布式数据接口

file system 文件系统

file attributes 文件属性

file format 文件格式

file functions 文件功能

file selection 文件选择

files in sub dir 子目录中的文件

file specification 文件标识，缩写为 file spec

filter 过滤器

find file 文件查寻

firewall 防火墙(是加强 Internet 与 Intranet(内部网)之间安全防范的一个系统)

firmware 固件

fixed disk 硬盘

fixed disk setup program 硬盘安装程序

fixes errors on the disk 解决磁盘错误

floppy disk 软盘

folder 文件夹

font 字体

form 格式

format diskette 格式化磁盘

formats a disk for use with MS-DOS 格式化用于 MS-DOS 的磁盘

form feed 进纸

fragments 分段

frame relay 帧中继

free memory 闲置内存

FTP(File Transfer Protocol) 文件传输协议

full screen 全屏

function 函数

## G

gateway 网关

GDI(Graphical Device Interface) 图形设备界面

global account 全局账号

global group 全局组

graphical 图解的

graphics library 图形库

graphics 图形

group account 组账号

group identifier 组标识符

group directories first 先显示目录组

GSNW  NetWare 网关服务

GUI(Graphical User Interface) 图形用户界面

## H

hard disk 硬盘

hardware detection 硬件检测

hash 散列表

help file 帮助文件

help index 帮助索引

help information 帮助信息

help path  帮助路径

help screen  帮助屏

help text  帮助说明

help topics  帮助主题

help window  帮助窗口

hidden file  隐含文件

hidden file attribute  隐含文件属性

home directory  主目录

homepage  主页

host  主机

HPFS(High-Performance File System)  高性能文件系统

HTML(HyperText Markup Language)  超文本标识语言

HTTP(HyperText Transport Protocol)  超文本传输协议

hyperlink  超级链接

hypertext  超文本

# I

ICMP(Internet Control Message Protocol)  网际控制报文协议

icon  图标

IE(Internet Explorer)  探索者(微软公司的网络浏览器)

IGMP(Internet Group Management Protocol)  Internet 群组管理协议

ignore case  忽略大小写

IGP(Interior Gateway Protocol)  内部网关协议

IIS(Internet Information Services)  信息服务器

image  图像

IMAP(Internet Message Access Protocol)  Internet 消息访问协议

impersonation attack  伪装攻击

in both conventional and upper memory  在常规和上位内存

incorrect DOS version  DOS 版本不正确

index server  索引服务器

indicates a binary file  表示是一个二进制文件

indicates an ASCII text file  表示是一个 ASCII 文本文件

inherited rights filter  继承权限过滤器

interactive user  交互性用户

interface  界面

intermediate system  中介系统

internal security  内部安全性

Internet server  因特网服务器

interpreter  解释程序

intranet  内联网，企业内部网

intruder  入侵者

in use  在使用

invalid directory  无效的目录

IP(Internet Protocol)  网际协议

IP address　IP 地址

IP masquerade　IP 伪装

IP spoofing　IP 欺骗

IPC(Inter-Process Communication)  进程间通信

IPX(Internet Packet eXchange)  互联网分组协议

IRQ(Interrupt Request)  中断请求

ISA(Industrial Standard Architecture)  工业标准结构总线

ISDN(Integrated Services Digital Network)  综合业务数字网

ISO(International Organization for Standardization)  国际标准化组织

ISP(Internet Service Provider)  互联网服务提供商

## J

jack in  一句黑客常用的口语，意思为破坏服务器安全的行为

JavaScript  基于 Java 语言的一种脚本语言

Java Virtual Machine　Java 虚拟机

# K

K bytes 千字节
kernel 内核
keyboard 键盘
keys 密钥
key space 密钥空间
keystroke recorder 按键记录器(一些用于窃取他人用户名和密码的工具)

# L

label disk 标注磁盘
LAN(Local Area Network) 局域网
LAN Server 局域网服务器
laptop 便携式电脑，笔记本电脑
largest executable program 最大可执行程序
largest memory block available 最大内存块可用
left handed 左手习惯
left margin 左边界
license 许可(证)
line number 行号
line spacing 行间距
list by files in sorted order 按指定顺序显示文件
list file 列表文件
local security 局部安全性
locate file 文件定位
log 日志、记录
logging 登录
logic bomb 逻辑炸弹(一种可导致系统加锁或者故障的程序或代码)
logical port 逻辑端口

logoff 退出、注销

logon script 登录脚本

logon 注册

lookup 查找

## M

Mac OS 苹果公司开发的操作系统(是一套运行于苹果 Macintosh 系列电脑上的操作系统)

macro name 宏名字

mainboard 主板

make directory 创建目录

manual 指南

MAPI(Mail Application Programming Interface) 邮件应用程序接口

mass browser 主浏览器

member server 成员服务器

memory info 内存信息

memory model 内存模式

menu bar 菜单条

menu command 菜单命令

message window 信息窗口

Microsoft corporation 微软公司

MIME(Multipurpose Internet Mail Extensions) 多媒体 Internet 邮件扩展

modem 调制解调器

module 模块

monitor mode 监控状态

monitor 监视器

monochrome monitor 单色监视器

mouse 鼠标

MPR(MultiProtocol Router) 多协议路由器

MUD(Multiple User Dimension; Multiple User Dialogue)一种通过网络让多人参与交谈式、探险式的角色扮演游戏

multi 多

multilink 多链接

multimedia 多媒体

multiprocessing 多重处理

## N

named pipes 命名管道

Navigator 引航者(网景公司的浏览器)

NDIS(Network Driver Interface Specification) 网络驱动程序接口规范

network layer 网络层

network monitor 一个网络监控程序

network operating system 网络操作系统

network printer 网络打印机

network security 网络安全

network user 网络用户

NFS(Network File System) 网络文件系统

NIC(Network Interface Card) 网卡(网络适配器)

NNTP(Network News Transport Protocol) 网络新闻传送协议

node 节点

## O

OA(Office Automation) 办公自动化

ODBC 开放数据库连接

online help 联机求助

OO(Object-Oriented) 面向对象

OpenGL(Open Graphics Library) 开放图形程序接口

option pack 功能补丁

optionally 可选择地

OS(Operation System) 操作系统

OSI Model 开放系统互连模式

OSPF(Open Shortest-Path First) 开放式最短路径优先协议

out-of-band attack 带外攻击

## P

packet filter 分组过滤器

page setup 页面设置

page frame 页面

page length 页长

pan 漫游

paragraph 段落

pseudo random 伪随机

password 口令

paste 粘贴

path 路径

pauses after each screen full of information 在显示每屏信息后暂停一下

payload 净负荷

PBX(Private Branch eXchange) 专用交换机

PCS(Personal Communications Service) 个人通信业务

PDC(Primary Domain Controller) 主域控制器

peer 对等

permission 权限

plaintext 明文

POP(Post Office Protocol) 互联网电子邮件协议标准

port 端口

POST(Power-On-Self-Test) 电源自检程序

postscript 附言

potential browser 潜在浏览器

P-P(Plug and Play) 即插即用

PPP 点到点协议

PPTP  点到点隧道协议

prefix to reverse order  反向显示的前缀

press a key to resume  按一键继续

press any key for file functions  按任意键执行文件功能

press esc to continue  按 esc 键继续

press esc to exit  按 esc 键退出

previous  前一个

print preview  打印预览

print all  全部打印

print device  打印设备

printer port  打印机端口

process  进程

processes files in all directories in the specified path  在指定的路径下处理所有目录下的文件

program  程序

program file  程序文件

programming environment  程序设计环境

prompts you before creating each destination file  在创建每个目标文件时提醒你

prompts you to press a key before copying  在拷贝前提示你按一下键

priority  优先权

protocol  协议

proxy server  代理服务器

proxy  代理

pull down  下拉

pull down menus  下拉式菜单

## Q

quick format  快速格式化

quick view  快速查看

# R

RAM(Random Access Memory) 随机存取内存，随机存取存储器

RAS(Remote Access Service) 远程访问服务

read only file 只读文件

read only mode 只读方式

redial 重拨

release 发布

remote boot 远程引导

remote control 远程控制

repeat last find 重复上次查找

replace 替换

report file 报表文件

resize 调整大小

restart 重新启动

right click 右击

right margin 右边距

RIP(Routing Information Protocol) 路由选择信息协议

ROM(Read Only Memory) 只读存储器

root directory 根目录

route 路由

routed daemon 一种利用 rip 的 Unix 寻径服务

router 路由器

routing table 路由表

routing 路由选择

row 列

RPC(Remote Procedure Call) 远程过程调用

RSA 一种公共密匙加密算法(RSA 公钥加密算法是 1977 年由 Ron Rivest、Adi Shamirh 和 LenAdleman 在美国麻省理工学院开发的；RSA 取自于他们三人的名字)

runtime error 运行时错误

## S

S/Key 安全连接的一次性密码系统(在 S/Key 中，密码从不会经过网络发送，因此不可能被窃取)

SACL(System Access Control List) 系统访问控制表

save all 全部保存

save as 另存为

scale 比例

scandisk 磁盘扫描程序

screen colors 屏幕色彩

screen options 屏幕任选项

screensaver 屏幕暂存器

screensavers 屏幕保护程序

screen size 屏幕大小

script 脚本

scrollbars 滚动条

scroll lock off 滚屏已锁定

SCSI(Small Computer System Interface) 小型计算机系统接口

search engine 搜索引擎

sectors per track 每道扇区数

secure 密码

select all 全选

select group 选定组

selection bar 选择栏

sender 发送者

server 服务器

server-based network 基于服务器的网络

service pack 服务补丁

session layer 会话层

set active partition 设置活动分区

settings 设置

setup 安装

setup options 安装选项

share/sharing 共享

share-level security 共享级安全性

shortcut 快捷方式

shortcut keys 快捷键

SID(Security Identifiers) 安全标识符

single side 单面

site 站点

SLIP(Serial Line Internet Protocol) 串行线网际协议

SMTP(Simple Mail Transfer Protocol) 简单邮件传送协议

sniffer 嗅探器(秘密捕获穿过网络的数据报文的程序,黑客一般用它来设法盗取用户名和密码)

SNMP(Simple Network Management Protocol) 简单网络管理协议

snooping 探听

sort order 顺序

specifies drive directory and or files to list 指定要列出的驱动器、目录和文件

specifies that you want to change to the parent directory 指定你想把父目录作为当前目录

specifies the file or files to be copied 指定要拷贝的文件

spoofing 电子欺骗(任何涉及假扮其他用户或主机以对目标进行未授权访问的过程)

SQL(Structured Query Language) 结构化查询语言

SSL(Secure Sockets Layer) 安全套接层

stack overflow 栈溢出

standalone server 独立服务器

startup options 启动选项

status bar 状态条

status line 状态行

step over 单步

stream cipher 流密码

strong cipher 强密码

strong password 强口令

style 样式

subdirectory 子目录

subnet mask 子网掩码

subnet 子网

swap file 交换文件

switches may be preset in the dircmd environment variable 开关可在 dircmd 环境变量中设置

sync 同步

system file 系统文件

system info 系统信息

# T

table 表

table of contents 目录

TCP/IP 传输控制协议/网际协议

telnet 远程登录

template 模版

terminal emulation 终端仿真

terminal settings 终端设置

test file 测试文件

TFTP(Trivial File Transfer Protocol) 普通文件传送协议

the active window 激活窗口

the two floppy disks must be the same type 两个软磁盘必须是同种类型的

thin client 瘦客户机

thread 线程

throughput 吞吐量

time bomb 时间炸弹(指等待某一特定时间或事件出现才激活,从而导致机器故障的程序)

toggle breakpoint 切换断点

tool bar 工具条

top margin 页面顶栏

trace route 一个 Unix 上的常用 TCP 程序,用于跟踪本机和远程主机之间的路由

transport layer 传输层

transport protocol 传输协议

Trojan horse 特洛伊木马

tunnel 安全加密链路

## U

UDP(User Datagram Protocol) 用户数据包协议

undo 撤销

uninstall 卸载

Unix 用于服务器的一种操作系统

unmark 取消标记

unselect 取消选择

update 更新

UPS(Uninterruptable Power Supply) 不间断电源

URL 统一资源定位器

USENET 世界性的新闻组网络系统

user account 用户账号

user name 用户名

uses lowercase 使用小写

uses wide list Format 使用宽行显示

## V

vector of attack 攻击向量

verifies that new files are written correctly 校验新文件是否正确写入了

video mode 显示方式

view window 内容浏览

view 视图

virtual directory 虚目录

virtual machine 虚拟机

virtual server 虚拟服务器

virus 病毒

vision 景象

vollabel 卷标

volume 文件集

volume label 卷标

volume serial number 卷序号

VRML(Virtual Reality Modeling Language) 虚拟现实建模语言

# W

WAN(Wide Area Network) 广域网

weak password 弱口令

web page 网页

website 网站

well-known ports 通用端口

Windows NT 微软公司的网络操作系统

wizzard 向导

word wrap 整字换行

working directory 正在工作的目录

workstation 工作站

worm 蠕虫

write mode 写方式

WWW(World Wide Web) 万维网

# X

X.25 一种分组交换网协议

XMS memory 扩充内存

## Z

zone transfer 区域转换

zoom in 放大

zoom out 缩小